Radicals

Radicals

D. C. NONHEBEL
Department of Pure and Applied Chemistry
University of Strathclyde

J. M. TEDDER
Purdie Professor of Chemistry
University of St Andrews

J. C. WALTON
Department of Chemistry
University of St Andrews

CAMBRIDGE UNIVERSITY PRESS
Cambridge
London · New York · Melbourne

Published by the Syndics of the Cambridge University Press
The Pitt Building, Trumpington Street, Cambridge CB2 1RP
Bentley House, 200 Euston Road, London NW1 2DB
32 East 57th Street, New York, NY 10022, USA
296 Beaconsfield Parade, Middle Park, Melbourne 3206, Australia

First published 1979

Typeset by H Charlesworth & Co Ltd, Huddersfield
Printed in Great Britain at the University Press, Cambridge

Library of Congress cataloguing in publication data
Nonhebel, D. C.
Radicals
(Cambridge texts in chemistry and biochemistry)
Includes index
1. Radicals (Chemistry) I. Tedder, John Michael, joint author.
II. Walton, John Christopher, joint author. III. Title.
QD471.N84   541'.224   78-54721
ISBN 0 521 22004 1
ISBN 0 521 29332 4 paperback

Contents

Foreword by Professor W. A. Waters   ix
Preface   xiii
Note on units   xv

1   Introduction   1
1.1.   Homolysis and heterolysis   1
1.2.   Electron spin and molecular orbitals   3
1.3.   Paramagnetism and electron spin   4
1.4.   The triplet state   4
1.5.   The reactions of radicals – the importance of chain processes   5
1.6.   The detection of radicals   7
1.7.   Reactivity, persistency and stability of radicals   8

2   The formation of radicals   10
2.1.   Thermolysis   10
2.2.   Photolysis   13
2.3.   Oxidation   18
2.4.   Reduction   18

3   Electron spin resonance spectroscopy   19
3.1.   Introduction   19
3.2.   Principles of ESR spectroscopy   19
3.3.   Characteristics of ESR spectra   21
3.4.   Spin trapping   28

4   Chemically induced dynamic nuclear polarization (CIDNP)   30
4.1.   The phenomenon   30
4.2.   The radical pair theory   33

5	**Stereochemistry of free radicals**	37
5.1.	Introduction	37
5.2.	Alkyl radicals	37
5.3.	Cycloalkyl radicals	39
5.4.	β-Substituted alkyl radicals	39
5.5.	Allylic radicals	42
5.6.	Vinylic radicals	42
5.7.	Group IV radicals	43
5.8.	Nitrogen-centred radicals	43
5.9.	Phosphorus-centred radicals	44
6	**Radicals of long life**	46
6.1.	Introduction	46
6.2.	Triarylmethyl radicals	46
6.3.	Phenoxy radicals	49
6.4.	Nitroxides	50
6.5.	Hydrazyl radicals	52
6.6.	Sterically hindered alkyl, aryl, vinyl and allyl radicals	53
7	**Radical – radical reactions**	55
7.1.	Introduction	55
7.2.	Atom combination reactions	56
7.3.	Combination and disproportionation of alkyl radicals	57
7.4.	Cage reactions and diffusion control in solution	59
7.5.	Combination and disproportionation of heteroradicals	60
8	**Radical transfer reactions**	62
8.1.	The kinetic problem	62
8.2.	The thermodynamic problem	64
8.3.	Substituent effects	68
8.4.	The vicinal effect, and the bridging atom hypothesis	75
8.5.	Allylic halogenation	76
8.6.	Solvent effects and reactions in solution	77
8.7.	Abstraction of atoms other than hydrogen	78
8.8.	Radical transfer reactions in industrial processes	80
9	**Combustion**	82
9.1.	The reaction of alkyl radicals with molecular oxygen	82
9.2.	Branching-chain reactions	83

9.3.	Explosions	84
9.4.	Slow oxidation of alkanes	86
10	**Radical addition reactions**	88
10.1.	Characteristic features of radical addition reactions	88
10.2.	Orientation of radical addition	90
10.3.	Carbon-carbon bond formation by radical addition	94
10.4.	Carbon-heteroatom bond formation by radical addition	97
10.5.	Radical fragmentation reactions	101
10.6.	Stereochemistry of addition	104
10.7.	Radical cyclization reactions	106
11	**Radical polymerization**	110
11.1.	Introduction	110
11.2.	Mechanism of radical polymerization	110
11.3.	Initiation	113
11.4.	Propagation	113
11.5.	Termination	114
11.6.	Chain transfer	116
11.7.	Copolymerization	117
11.8.	Principles of polymer synthesis	121
12	**Homolytic aromatic substitution**	123
12.1.	Introduction	123
12.2.	Mechanism of homolytic aromatic substitution	123
12.3.	Arylation	125
12.4.	Acyloxylation	130
12.5.	Alkylation	130
12.6.	Hydroxylation	131
12.7.	Homolytic substitutions of polycyclic aromatic hydrocarbons	132
12.8.	Homolytic substitutions of heteroaromatic compounds	133
13	**Radical oxidations and reductions**	135
13.1.	Introduction	135
13.2.	Oxidation of radicals	136
13.3.	Reduction of radicals	138
13.4.	Mechanisms for oxidation of organic compounds with metal-ion oxidants	138
13.5.	Kolbe oxidation of carboxylic acids	140
13.6.	Oxidation of phenols	141

13.7.	Metal-catalysed reductions of organic compounds	144
13.8.	Redox reactions	145
13.9.	Electron-transfer substitution reactions	147
14	**Autoxidation**	150
14.1.	Introduction	150
14.2.	Kinetic features of autoxidation	151
14.3.	Catalysis of autoxidation by metal salts	154
14.4.	Antioxidants	155
14.5.	Autoxidation of aldehydes	157
14.6.	Autoxidation of alkenes	158
15	**Radical rearrangements**	160
15.1.	Introduction	160
15.2.	Aryl migrations	161
15.3.	1,2-vinyl migrations	163
15.4.	Ring opening of cyclopropyl radicals	164
15.5.	Migrations of hydrogen	164
15.6.	The Barton reaction	166
15.7.	Molecular rearrangements involving a radical-pair mechanism	167
16	**Radicals in biological systems**	170
16.1.	Introduction	170
16.2.	Biological oxidations	170
16.3.	Autoxidation in biologically important compounds	175
16.4.	Mechanism of action of coenzyme B_{12}	179
16.5.	Spin labels	182
17	**Radical displacement reactions**	185
17.1.	Nature of the reaction	185
17.2.	Displacement at group II and group III elements	186
17.3.	Displacement at group IV elements	187
17.4.	Displacement at group V elements	187
17.5.	Displacement at group VI elements	188
17.6.	Displacement at mercury and transition metal atoms	189
17.7.	Stereochemistry of homolytic displacement reactions	190
	Suggestions for further reading	192
	Index	198

Foreword

By W. A. Waters, FRS, Emeritus Professor of Chemistry, Oxford University

It is little more than 40 years since D. H. Hey and I independently suggested, in this country, that a few chemical reactions in the liquid phase might involve the formation of short-lived free radicals such as $C_6H_5\cdot$, the neutral phenyl radical. At first these views were treated with scepticism since the then current theories of 'chemical polarity' indicated that, if organic molecules dissociated at all during the courses of their reactions in solution, then they decomposed to give charged ions and not electrically neutral fragments such as free atoms of bromine or simple compounds of trivalent carbon. These had been shown nearly a century before to combine at once to dimeric molecules – e.g. $2Na + 2CH_3Br \rightarrow C_2H_6 + 2NaBr$ – whenever attempts had been made to prepare them. In general, organic chemists thought that additions and substitutions were both single-step reactions and that there was no reason to postulate that active short-lived intermediates might be involved except perhaps in surface-catalysed processes.

Within ten years however, largely due to outstanding work by the late M. S. Kharasch and his colleagues, the occurrence of free-radical reactions in solution was clearly recognized and many were being studied intensively in the USA where, stimulated by a war-time need for synthetic rubber, it was shown that petroleum was not merely a convenient fuel but a more valuable source of desirable synthetic chemicals than coal tar. During the same period the acceptance of the kinetic theories of E. D. Hughes and C. K. Ingold for reactions such as hydrolysis showed the fundamental importance in chemistry of establishing not merely what molecules could interact but how the reaction processes occur.

Today it is so well recognized that during chemical reactions specific covalent, i.e. electron-pair, bonds in compounds are broken either symmetrically by *homolysis* to give free radicals or unsymmetrically by *heterolysis*, that a basic understanding of the physical and chemical nature of free radicals is a requisite for all students who wish to gain more than a rudimentary knowledge of organic chemistry. This concise book is intended for undergraduate students who are nearing their degree examinations and employs the mechanistic approach to the subject. Merely by glancing at the chapter titles one can see the great extent to which the concept of the transient existence of free radicals during the course of reactions, such as addition, substitution, polymerization, oxidation and reduction has now permeated the whole of organic chemistry. Readers of the book can notice too that the study of free radicals has brought about a welcome unification of theories apposite to the physical, organic and inorganic branches of chemical science.

In the early days of free-radical chemistry no physical methods were known for the specific detection of short-lived uncharged particles formed during chemical reactions. Evidence for the participation of free radicals in certain reactions in solution came from the kinetic similarity of these reactions with gas phase reactions in which free atoms and free radicals were involved. In particular, rate equations diagnostic of chain reactions which had been formulated early in this century by physical chemists for such simple overall reactions as $H_2 + Cl_2 = 2HCl$ were proved to be equally applicable to the peroxide-catalysed addition of hydrogen bromide to propylene and to the peroxide-catalysed polymerizations of vinyl chloride, styrene, methyl methacrylate and other olefinic compounds to industrially important products.

The study of free-radical reactions has also broken down other artificial distinctions between physical and organic chemistry. For instance *photochemistry*, originally concerned only with the absorption or emission of radiant energy by simple molecules has now become an integral part of free-radical chemistry. As a technique for the homolysis of specifiable covalent bonds photolysis receives mention in almost every chapter of this book.

Again within the past twenty years the discovery by physicists of the technique of *electron spin* (or *paramagnetic*) *resonance* (ESR or EPR) has provided unequivocal proofs of the independent existence of both long-lived and extremely short-lived free radicals in reacting systems, since ESR spectra are produced only by particles which contain an unpaired electron. This physical technique has verified all the early hypotheses of the participation of free radicals in certain chemical reactions. Moreover, since ESR spectra can be used for the diagnosis of radical structure it has helped greatly in the recognition of new types of free radicals and of new homolytic reactions.

Among these newer studies those involving oxidations and reduction are noteworthy. It is now evident that most of the redox reactions of the simple or complexed ions of the transition metals, such as iron, copper and cobalt, proceed by the process of one-electron transfer and thereby generate free radicals. Free-radical reactions consequently are now important in inorganic chemistry, which has been further revitalized by the discovery and study of free radicals containing unpaired electrons on atoms of many other elements throughout the periodic system.

Finally, as chapter 16 of this book indicates, the study of free radicals is now having an impact on biochemistry. The long-known facts (i) that molecular oxygen in its most stable state is a biradical, and (ii) that a number of transition metals such as iron and copper are essential 'trace elements' for the healthy life of both plants and animals, lead one to infer that homolytic reactions must somehow be involved in both the oxidative metabolism of foodstuffs by animals and the reduction of carbon dioxide to plant tissues by photosynthesis. Mechanistic organic chemistry is now pervading biochemistry and in coming years the consideration of free-radical reactions is, I feel sure, going to make a great impact on enzymology.

By looking at the contents of this book as a whole and noting the great diversification of the aspects of chemistry with which it is now concerned I cannot but feel that the progress of free-radical chemistry during the past forty years resembles that of a branched chain reaction. It has stimulated experimental discoveries, as in an explosion, that cannot be confined to the boundary walls of any academic lecture course or syllabus. Nor can, or should, its implications be definable within any computerizable library scheme.

Preface

We have written this book with the intention of showing students something of the importance and wide occurrence of reactions involving one-electron transfer. The aim of the book is to provide coverage sufficient for the non-specialist student, while also providing an adequate introduction for those who wish to know more about radical reactions (further reading is suggested at the end of the book). It is assumed that the reader will already have had some grounding in basic physical and organic chemistry and will be familiar with the common spectroscopic techniques (IR, UV and NMR). The book is primarily intended for students reading chemistry as their main subject for use in the latter part of their course.

In our opinion it is most important that students of chemistry should be familiar with the wide range and practical importance of radical reactions; we have therefore included chapters on combustion, polymerization and autoxidation. The importance of one-electron processes in biological systems is becoming increasingly apparent and chapter 16 introduces this field. Spectroscopic techniques are vital in many radical studies, and ESR and the more recently developed CIDNP are introduced for the non-specialist in chapters 3 and 4. We believe that a study of radical reactions provides the student with a good experimental background in which to develop his theoretical ideas. We hope this book may stimulate interest in both applied and theoretical aspects of chemistry.

We express our thanks firstly to our students and research students whose interest in radical chemistry has provided the *raison d'être* of this volume. We thank many colleagues who have read parts of the manuscript and made useful comments and would mention in particular Professor P. L. Pauson, Drs A. R. Butler, I. R. Dunkin, K. U. Ingold, A. Nechvatal and C. J. Suckling. We acknowledge the kind permission of Professors R. Kaptein, R. O. C. Norman, FRS, M. J. Perkins, H. R. Ward and W. A. Waters, FRS, to reproduce diagrams. We also thank Mrs A. Cumming and Mrs B. Stewart for typing, retyping (and re-retyping) the manuscript, and Mrs J. Walton for reading and correcting proofs. We particularly wish to thank Professor K. Schofield for the careful and extremely thorough editorial work. Finally we thank Professor W. A. Waters, FRS, for reading the whole manuscript, making many valuable suggestions and for writing the Foreword.

D. C. Nonhebel
J. M. Tedder
J. C. Walton

Note on units

We have written this book at a time when units and the symbols for them are being changed. A new international system is gradually being adopted. Since we are in a period of transition we have used a mixture of SI units and the previously accepted units. The units we have used most frequently, and their relation to other common units, are tabulated below.

Thermodynamic data

Energies in joules (J) (1 calorie = 4.184 J)
Heats of reaction (ΔH) in kilojoules mole^{-1} (kJ mol^{-1})
Entropies (S) in joules degree^{-1} (J K^{-1})
Entropies of reaction (ΔS) in joules degree^{-1} mole^{-1} (J K^{-1} mol^{-1})
Ionization potentials in electron volts (eV)
Pressure in torr (1 torr = 1 mm Hg)

Kinetic data

Rate constants of first-order reactions in second^{-1} (s^{-1})
Rate constants of second-order reactions in litres mole^{-1} second^{-1} (l mol^{-1} s^{-1})
Activation energies in kilojoules mole^{-1} (kJ mol^{-1})
A-factors have the same units as the rate constants of the reactions from which they are derived.

Ultraviolet spectra

Wavelength in nanometres (nm = 10^{-9} m)

Infrared spectra

Wavenumber in centimetre^{-1} (cm^{-1})
Force constants in millidynes per ångstrom (mdyn Å^{-1})

Nuclear magnetic resonance spectra

Chemical shifts in parts per million (ppm)
Operating frequency in hertz (Hz)

Electron spin resonance spectra

Magnetic induction in tesla
Coupling constants in millitesla (1 mT = 10 gauss)

Bond lengths

Lengths in ångstroms (1 Å = 10^{-1} nm)

1 Introduction

1.1. Homolysis and heterolysis

Before we can consider the nature of 'radicals' we must first understand something about the nature of a chemical bond. We must understand why the two atoms in a hydrogen molecule stay bound together. The reason is that this is a lower energy (potential energy) state than two separate hydrogen atoms. If we supply sufficient energy to a hydrogen molecule we can raise its energy above that of two hydrogen atoms and if we do this the molecule will dissociate into its constituent atoms. The energy we have to put into the molecule, i.e. the energy required to break the hydrogen–hydrogen bond is called the *bond dissociation energy* and is written $D(H-H)$. If two isolated hydrogen atoms come together to form a bond exactly this amount of energy will be released. If the new molecule is isolated in free space and is unable to transmit part of this energy to another species (a so-called '*third body*') the atoms will simply dissociate again. We shall see later that the combination of two atoms to form a molecule is always pressure-dependent in the gas phase, because of the need of a third body to absorb part of the energy.

Two atoms come together to form a molecule because the potential energy will be less. According to all modern theories of valency the bond between two hydrogen atoms forming a hydrogen molecule is due to the sharing of their two electrons. In other words the energy of two electrons together associated with the two nuclei is lower than the energy of the two electrons separately associated with one nucleus each. When two atoms share two electrons so that the electrons 'bind' the atoms together, we call this a single bond. As the term single bond implies, we can also have double bonds in which two atoms share four electrons, triple bonds in which two atoms share six electrons and we can also have bonds of partial order. At present we are concerned with single bonds. It is clear that a single bond between two different atoms A and B can break in three different ways:

(1) atom A can take both electrons from the bond, and remembering electrons

are negatively charged, we have:

$$A : B \rightarrow A :^- \quad B^+$$

(2) atom B can take both electrons

$$A : B \rightarrow A^+ \quad : B^-$$

(3) each atom can take one of the bonding electrons

$$A : B \rightarrow A \cdot \quad \cdot B$$

The charged species formed in (1) and (2) are called *ions* and the fission of a single bond to form ions is called *heterolysis*. The uncharged species formed in (3) are called *radicals* and the fission of a single bond into two radicals is called *homolysis*. *The lowest energy pathway for the fission of a single bond is always a homolysis.* Even sodium chloride when heated in *the gas phase* will dissociate into sodium atoms and chlorine atoms. In solution, especially in a solution of high dielectric constant, sodium chloride ionizes into sodium ions (Na^+) and chloride ions (Cl^-); this is because in solution the ions become surrounded by a shell of solvent molecules. The interaction between the solvent and charged species, called *solvation,* greatly lowers the potential energy and is a major controlling factor of all ionic reactions in solution. Solvation can also be important where no actual ions are formed but polar molecules or intermediates are involved in a reaction. The most obvious example of solvation is the dissociation of compounds like sodium chloride into ions in solution. So tightly do the solvent shells become bound round the ions that, in spite of the strong electrostatic attraction between them, the sodium ions do not immediately combine with the chloride ions.

Radicals are not solvated in the same way as ions. Ionic solvation involves electron-pair interaction. To solvate radicals, a solvent containing unpaired electrons is necessary. In general, solvents have all their electrons paired (see below) so that the interaction between a radical and a solvent is very weak. This means that there is no strong solvent shell round a radical so that if two radicals come together they combine to form a molecule.

It is important to distinguish clearly between ions and radicals; both can formally be considered as derived from the fission of a single bond. Fission of a single bond in the gas phase will yield radicals rather than ions. In solution polar bonds can undergo heterolysis to form ions at low temperatures, while weak non-polar bonds will undergo homolysis to yield radicals usually only at elevated temperatures. The ions surrounded by their solvent shells will not recombine in spite of their electrostatic attraction, while radicals will recombine almost on every collision because they are not surrounded by the solvent shell. Thus it is possible to have very high concentrations indeed of ions in polar solvents like water. On the other hand we can never have high concentrations of common radicals. There is a special class of free radicals called *'persistent radicals'* (see chapter 6) which can exist as pure liquids or even in the solid state. It is important to appreciate this is a special class and in general radicals are transient species with very short lifetimes. A solvated fluoride ion can exist in aqueous solution almost indefinitely; a fluorine atom, which cannot be solvated, reacts extremely rapidly with almost every common solvent from water to petrol.

1.2. Electron spin and molecular orbitals

According to the concepts of quantum mechanics atoms are made up of positively charged nuclei surrounded by negatively charged electrons. The electrons occupy atomic orbitals each of which is associated with a particular energy. A neutral atom contains sufficient electrons to balance the nuclear charge. Thus a carbon nucleus carries a charge of +6 (in atomic units) and in a neutral atom this is surrounded by six electrons occupying four atomic orbitals. The electrons go into the orbitals of lowest energy available in pairs with antiparallel spins. Electrons can take up two kinds of spin and two electrons can either have their spins parallel or antiparallel. When there are two or more orbitals with the same energy (called *'degenerate orbitals'*) and only two electrons available to go into them, the electrons preferentially go one into each orbital with parallel spins (*Hund's rule*).

If we can have atomic orbitals, then it is reasonable to expect molecular orbitals and we find that electrons in the highest filled atomic orbitals move into molecular orbitals when a bond is formed. It is not the purpose of this book to discuss valency theories in detail, but there are important features of molecular orbitals we must take note of. According to our discussion so far, when two hydrogen atoms come together to form a molecule their electrons (originally in 1 s atomic orbitals) move into a molecular orbital which encompasses both atoms and in which the electrons have antiparallel spins.

An atom has s-orbitals which are spherically symmetric around the atomic nucleus and p-orbitals which have a nodal plane containing the nucleus (we need not concern ourselves with d- and f-atomic orbitals). In a similar way diatomic molecules have σ-orbitals which are cylindrically symmetric around an axis joining the two atoms, and π-orbitals which have a nodal plane containing both atoms. For every bonding molecular orbital, there is a corresponding antibonding orbital. This is a high-energy state and in a simple diatomic molecule represents a repulsive state. The antibonding orbitals corresponding to the σ- and π-orbitals are designated σ* and π*. Just as in a many-electron atom a $2p$ atomic orbital is of higher energy than a $2s$ atomic orbital, so a π molecular orbital is of higher energy than a σ-orbital. The reverse however applies to the antibonding orbitals, a σ*-orbital is of higher energy than a π*.

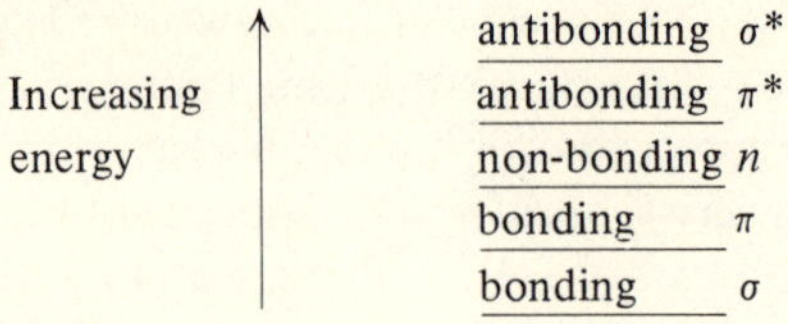

	antibonding σ^*
Increasing	antibonding π^*
energy	non-bonding n
	bonding π
	bonding σ

Figure 1.1. Energy states of molecular orbitals.

1.3. Paramagnetism and electron spin

If two magnetic poles p_1 and p_2 are separated by a distance r in a medium X, the force F between the poles is given by

$$F = \frac{1}{\mu_X} \frac{p_1 p_2}{r^2}$$

The quantity μ_X is the magnetic permeability of substance X (the magnetic analogue of electrical permittivity). The magnetic susceptibility χ_X is given by $\chi_X = \mu_X/\mu_0$ (where μ_0 is the permeability of a vacuum). Substances for which χ_X is slightly less than one are called *diamagnetic*, substances for which χ_X is slightly greater than one are called *paramagnetic* and substances for which χ_X is very large are called *ferromagnetic*. A paramagnetic or ferromagnetic substance is drawn into a strong magnetic field whereas a diamagnetic one is pushed out. Diamagnetism is associated with all substances. Electrons moving around nuclei form electric currents, and if a magnetic field is applied the velocity of the electrons is changed inducing a magnetic field which is in opposition to the applied field. Diamagnetic susceptibility is therefore always negative and diamagnetic substances are pushed out of the field. Paramagnetism arises from permanent molecular dipole moments which are partly due to the magnetic moments associated with the orbital movements of unpaired electrons and partly due to the spin magnetic moments of unpaired electrons. In organic molecules there are strong internal electric fields which prevent the unpaired electrons lining up with an external field, and the angular momentum component is said to be *quenched*. Only the effect due to electron spin remains (nucleon spin encountered in NMR spectroscopy is less than 10^{-3} of that due to the electron). For the study of radicals, magnetic susceptibility experiments have been superseded by paramagnetic resonance spectroscopy or as it is commonly called, electron spin resonance spectroscopy (ESR). This is discussed in detail in chapter 3 and can tell us a great deal about the structure of radicals.

1.4. The triplet state

There are some simple molecules which are radicals, i.e. they have unpaired electrons. The most obvious examples are nitric oxide and nitrogen dioxide.

Both these molecules have odd numbers of electrons which means they are radicals. Nitric oxide shows very little tendency to dimerize in the gas phase, but nitrogen dioxide is in equilibrium with its dimer, dinitrogen tetroxide. Both these compounds are paramagnetic as we would expect for any species with an odd number of electrons. However molecular oxygen which has an even number of electrons is also paramagnetic. To understand the paramagnetism of oxygen we must look at the molecular orbitals of the oxygen molecule (figure 1.2).

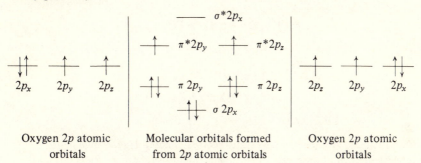

| Oxygen 2p atomic orbitals | Molecular orbitals formed from 2p atomic orbitals | Oxygen 2p atomic orbitals |

Figure 1.2. Molecular orbital diagram for molecular oxygen.

The diagram shows eight of the valence electrons in an oxygen molecule; these successively fill a σ-orbital and two π-orbitals (not shown are the 2s atomic orbitals which combine to give two σ-orbitals). We are then left with two electrons which according to Hund's rule must go one each into the two degenerate π^*-orbitals with parallel spins. It is the presence of these unpaired electrons which causes the paramagnetism of molecular oxygen. The unpaired electrons also control the chemistry of oxygen which behaves like a stable biradical and, particularly in chapter 9, we shall have much to say about its reactions with other radicals. Molecular species which have two electrons with parallel spins are called *triplets*, the name arising from spectroscopy. Molecules, like oxygen, which have a triplet ground state are rare, but we shall see (chapter 2) that electronically excited molecules are often triplets.

1.5.   The reactions of radicals – the importance of chain processes

We have seen that, with the exception of 'persistent radicals' (to be discussed in chapter 6), when two radicals come together they usually combine or disproportionate. We have also seen that, unlike ions, radicals in solution are not surrounded by a solvent shell, so that whereas we can have very high concentrations of ions, high concentrations of radicals (excluding persistent radicals) are impossible. In general, radicals are extremely reactive species reacting rapidly with the majority of organic molecules, including the alkanes which are normally completely resistant to the action of ions.

We can list some of the main reactions of radicals as follows:

$$X \cdot + Y \cdot \rightarrow X—Y \qquad\qquad \textit{combination}$$

$$(\text{e.g. } CH_3 \cdot + CH_3 \cdot \rightarrow C_2 H_6)$$

$$X \cdot + Y—Z—W \cdot \rightarrow XY + Z=W \qquad \textit{disproportionation}$$

$$(\text{e.g. } CH_3 \cdot + CH_3 CH_2 \cdot \rightarrow CH_4 + CH_2=CH_2)$$

$$Y—Z \cdot \rightarrow Y \cdot + Z \qquad\qquad \textit{fragmentation}$$

$$(\text{e.g. } CH_3 CO \cdot \rightarrow CH_3 \cdot + CO)$$

$$X \cdot + Y—Z \rightarrow X—Y + Z \cdot \qquad \textit{radical transfer*}$$

$$(\text{e.g. } Cl \cdot + CH_4 \rightarrow HCl + CH_3 \cdot)$$

$$X \cdot + Y=Z \rightarrow X—Y—Z \cdot \qquad\qquad \textit{addition}$$

$$(\text{e.g. } CF_3 \cdot + CH_2=CH_2 \rightarrow CF_3 CH_2 \overset{\cdot}{C}H_2)$$

(Not included are one-electron transfers (see chapter 2) rearrangements (chapter 15) and homolytic substitution reactions (chapter 17).)

Notice that only in the radical–radical reactions, combination and disproportionation (see chapter 7), do we obtain non-radical products. Radical-radical reactions are usually extremely fast, but they depend on the square of the radical concentration. We have already emphasized that normally the radical concentration is extremely low so that radical transfer and addition reactions predominate and these usually involve a chain process e.g.

$$Cl_2 \overset{h\nu}{\rightarrow} 2Cl \cdot \qquad\qquad \textit{initiation}$$

$$\left. \begin{array}{l} Cl \cdot + CH_4 \rightarrow HCl + CH_3 \cdot \\[6pt] CH_3 \cdot + Cl_2 \rightarrow CH_3 Cl + Cl \cdot \end{array} \right\} \quad \textit{chain propagation}$$

$$\left. \begin{array}{l} Cl \cdot + Cl \cdot + M \rightarrow Cl_2 + M \\[6pt] Cl \cdot + CH_3 \cdot \rightarrow CH_3 Cl \\[6pt] CH_3 \cdot + CH_3 \cdot \rightarrow C_2 H_6 \end{array} \right\} \quad \textit{chain termination}$$

All radical reactions require initiation and this will be discussed in the next chapter. The two *chain propagation* steps in the above sequence are radical transfer reactions (see chapter 8) and clearly if the termination steps are absent these reactions could continue until all the reagents were consumed. Chlorination of this kind involves particularly long chains (of the order of 10^6 cycles), but reaction chain lengths can vary and in many cases only a very few

---

* When Y is a group rather than an atom, this reaction is referred to as *substitution*.

cycles occur. Chain termination processes will be discussed in chapter 7. Addition reactions also usually involve chain processes which can lead to simple adducts (see chapter 10) when an addition process is succeeded by a radical transfer:

$$CCl_3\cdot + CH_2 = CH_2 \xrightarrow{addition} CCl_3CH_2CH_2\cdot$$

$$CCl_3CH_2CH_2\cdot + CCl_3Br \xrightarrow[transfer]{radical} CCl_3CH_2CH_2Br + CCl_3\cdot$$

$\left.\right\}$ *chain propagation*

or by polymerization (see chapter 10) when one addition step is followed by another:

$$CF_3\cdot + CF_2 = CF_2 \xrightarrow{addition} CF_3CF_2CF_2\cdot$$

$$CF_3CF_2CF_2\cdot + CF_2 = CF_2 \xrightarrow{addition} CF_3(CF_2)_4\cdot$$

$\left.\right\}$ *chain propagation*

Chain reactions are indeed one characteristic of radical chemistry and chapters 9–11 will be much concerned with reactions of this type.

1.6.   The detection of radicals

As we shall see in chapter 6 there is a group of persistent radicals which can be obtained in high concentration in solution, or even as pure liquids and solids. Such species can be studied by all the conventional techniques available to the chemist, their infrared and ultraviolet spectra can be determined and direct magnetic susceptibility experiments confirming the presence of unpaired electrons are possible. More important their ESR spectra can be studied and a great deal about their structure can be deduced from these (see chapter 4).

The majority of radicals exist only as transient intermediates, which are never normally present in any large concentration. The occurrence of radical intermediates can often be inferred from the nature of the reaction products or from a study of the kinetics. Alternatively the reaction may be inhibited by persistent radicals such as nitroxides or galvinoxyl which trap the transient intermediate radical. However confirmation of radical intermediates by physical methods usually involves generating a high concentration of radicals by irradiation with an intense light source, or preparing the radicals by electron transfer from a transition metal ion which can be present in high concentration. Flash photolysis has enabled the electronic spectra of many radicals to be studied. Others have been studied in the ESR spectrometer when very intense light sources have been used. One method is to dissolve the compound to be photolysed in a mixed solvent which on cooling forms a glass. On photolysis the radicals are trapped in the glass and their spectra can be observed. Another technique is to mix a solution containing a high concentration of the ions of a transition metal with one containing a high concentration of the anions to be oxidized to radicals just before passage through the cavity of the ESR spectrometer. Radicals can sometimes be detected by mass spectrometric techniques, but very special conditions are required.

Radical and ionic reactions can occur together, particularly in addition processes. Usually radical processes are favoured in the gas phase or in non-polar media, whereas ionic reactions are favoured in solvents of high dielectric constant.

1.7. Reactivity, persistency and stability of radicals

Most radicals exist only as transient intermediates occurring during a reaction process. There are however radicals which exist as pure liquids or stable solids. These latter radicals used to be called 'stable radicals', but this term is undesirable because a highly reactive methyl radical is perfectly stable in isolation. It has therefore become usual to refer to the unreactive radicals which can exist in high concentration in solution without either reacting with the solvent or combining with each other, as 'radicals of long life' or 'persistent radicals' and the chemistry of these species is discussed in chapter 6.

The radicals of short life exhibit a wide range of reactivities as determined by the readiness with which they undergo radical transfer or radical addition reactions, though their combination rates will probably be similar, and in solution will be diffusion controlled. However it is very difficult to find a scale of reactivity which can be regarded as standard since we shall see in subsequent chapters that the relative order of reactivity of a pair of radicals attacking one site may be reversed in their attack at another site. An order of intrinsic reactivity could be defined by the strength of the carbon–hydrogen bond at the radical centre e.g.

$$C_6H_5\cdot, CH_2=CH\cdot > CH_3\cdot > RCH_2\cdot > R_2CH\cdot > R_3C\cdot > C_6H_5CH_2\cdot,$$

$$CH_2=CHCH_2\cdot \quad \text{(where R = alkyl)}$$

A similar order of intrinsic reactivity can be derived from the relative rates of decomposition of the symmetrical azo-compounds.

$$R-N=N-R \rightarrow 2R\cdot + N_2$$

The analogous decomposition of symmetrical peroxides will not give the same order of reactivity because polar effects contribute, and this emphasizes the difficulty of defining a uniform scale.

Just as carbonium ions are stabilized by delocalization of the charge through a π-orbital system (e.g. benzyl) so a similar stabilization of a radical is provided by delocalization of the unpaired electron. However the stability provided by delocalization (resonance stabilization) is much less important in radicals than in carbonium ions. The stability of the triphenyl carbonium ion is largely due to the delocalization of the positive charge over the *ortho-* and *para-*positions of the benzene rings. The persistency of the triphenylmethyl radical on the other hand is due primarily to steric effects and only slightly attributable to delocalization. Nonetheless delocalization of the unpaired electron is a factor in determining the reactivity of a radical even if it is often outweighed by other factors such as steric compression and polarity.

2   The formation of radicals

There are two ways in which radicals can be formed: (i) by the fission of a single bond and (ii) by a one-electron transfer from or to an ion.

$$A-B \rightarrow A\cdot \quad B\cdot \qquad \textit{bond fission}$$

$$A:^- - e^- \rightarrow A\cdot$$
$$A^+ + e^- \rightarrow A\cdot$$

$\qquad\qquad\qquad\qquad$ *one-electron transfers*

The simplest example of bond fission is the mechanical fracture of any polymeric material. If a piece of thread (nylon or cotton) or a rubber band is stretched until it breaks, radicals will be found on either side of the fracture. Such a process has no practical value, although mechanical formation of radicals can be important in the degradation of polymers. In the laboratory the energy required to break single bonds is usually either supplied thermally or photochemically and the two processes are known as *thermolysis* and *photolysis*. One-electron transfer can involve inorganic ions or electrochemical processes; the donation of an electron is a *reduction* and the abstraction of an electron is an *oxidation*.

2.1.   Thermolysis

If the temperature is raised sufficiently all chemical bonds can be broken. In the introduction we observed that two atoms joined together to form a molecule because this is a state of lower energy, and that the energy required to dissociate the molecule into its constituent atoms was called the bond dissociation energy. In a polyatomic molecule different bonds will have different energies, e.g. in bromomethane the carbon–bromine bond is considerably weaker than the carbon–hydrogen bonds (i.e. $D(H_3C-Br) < D(BrH_2C-H)$). It is important to distinguish between *bond dissociation energy* which refers to a particular bond, and simple *bond energy* which is an average value computed for the whole molecule. The distinction is important because the energy required to break one of the C—H bonds in methane is not the same as the energy required to break a C—H bond in the methyl radical.

Two atoms joined together by a single bond vibrate like two weights attached to each other by a spring. The frequency of vibration can only have certain fixed values, i.e. the vibration is quantized. Just as only certain electron levels are possible, so only specific vibration levels are allowed. The absorption bands observed in the infrared are due to transition from one vibrational level to the next. In a normal diatomic molecule there are several vibrational levels and at low temperatures most molecules are in the lowest level. As the temperature rises so a molecule attains the higher vibrational levels until eventually the vibrational energy exceeds the bond dissociation energy and the molecule breaks up.

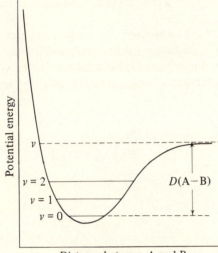

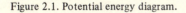
Distance between A and B

Figure 2.1. Potential energy diagram.

A very important question, which has been the object of much study is 'How does the energy reach a molecule in the gas phase?' Experiment shows that a molecule acquires energy by collision. The energized molecule can either dissociate or lose its energy also by collision. We can write the following simple mechanism:

$$A + A \rightleftharpoons A^* + A$$

$$A^* \rightarrow \text{dissociated products}$$

(* indicates an excited molecule)

At high pressures collision between molecules will be very frequent; the rate-determining step will be the dissociation of the excited molecule A* and the reaction will be first order, the rate depending on the overall concentration of A. As the pressure drops the reaction order will change, collisions will become less frequent and eventually the rate-determining step will be the rate of formation of the energized molecule A*; this is a second-order reaction, the rate depending on the square of the overall concentration of A. Full theoretical treatment of unimolecular reactions is much more complicated than we have presented here, and its study has been of great importance. In solution the dissociating molecule can gain its energy from collisions with solvent molecules and the reaction is first order for all concentrations. However a new problem now arises, the dissociating fragments will be surrounded by solvent molecules and they are therefore likely to undergo many collisions which may lead to recombination before they diffuse away from each other through the solvent. This 'cage effect' will be discussed again in later chapters especially chapter 7.

Although all chemical bonds will be broken if the temperature is raised sufficiently, certain bonds are particularly weak and compounds with such bonds are used to initiate radical reactions. Typical weak bonds are oxygen–oxygen bonds in peroxides and carbon–nitrogen bonds in azo-compounds.

$$
\underset{\underset{CH_3}{|}}{\overset{\overset{CH_3}{|}}{CH_3-C}}-O-O-\underset{\underset{CH_3}{|}}{\overset{\overset{CH_3}{|}}{C}}-CH_3 \xrightarrow{\Delta} 2\ CH_3-\underset{\underset{CH_3}{|}}{\overset{\overset{CH_3}{|}}{C}}-O\cdot
$$

Di-t-butylperoxide

$$
\underset{\underset{CN}{|}}{\overset{\overset{CH_3}{|}}{CH_3-C}}-N=N-\underset{\underset{CN}{|}}{\overset{\overset{CH_3}{|}}{C}}-CH_3 \xrightarrow{\Delta} 2\ CH_3-\underset{\underset{CN}{|}}{\overset{\overset{CH_3}{|}}{C}}\cdot + N_2
$$

Azobisisobutyronitrile (AIBN)

The halogen molecules will dissociate into atoms at comparatively low temperatures, especially fluorine which is very slightly dissociated even at room temperature.

$$F_2 + M \rightleftharpoons 2F\cdot + M$$

$$(K_p^{298} = 2.02 \times 10^{-16}\ N\,m^{-2})$$

The important feature of fluorine and to a lesser extent of chlorine, is that even though the atom concentration is very low the atoms are so reactive that a chain process can be initiated.

The compounds with weak bonds like the peroxides and azo-compounds are very important as *initiators*. The peroxide is used to initiate a chain reaction, the initiator only being present in very small amount.

$$C_6H_5COO-OOCC_6H_5 \rightarrow 2C_6H_5CO_2 \cdot$$

$$C_6H_5CO_2\cdot + C_3H_7CHO \rightarrow C_3H_7\dot{C}O + C_6H_5CO_2H$$

⎱ *initiation*

$$C_3H_7\dot{C}O + CH_2=CHR \rightarrow C_3H_7COCH_2\dot{C}HR$$

$$C_3H_7COCH_2\dot{C}HR + C_3H_7CHO \rightarrow C_3H_7COCH_2CH_2R + C_3H_7\dot{C}O$$

⎱ *chain propagation*

The thermal decompositions of diacyl peroxides such as dibenzoyl peroxide do not show simple first-order kinetics but exhibit a rate law $-d[P]/dt = k_d[P] + k_i[P^{1/2}]$ (where $[P]$ corresponds to peroxide concentration). This can be explained by a mechanism of the type shown

$$(C_6H_5CO_2)_2 \rightarrow 2C_6H_5CO_2\cdot$$

$$C_6H_5CO_2\cdot + ArH \rightarrow C_6H_5CO_2\dot{A}rH$$

$$C_6H_5CO_2\dot{A}rH + (C_6H_5CO_2)_2 \rightarrow C_6H_5CO_2\cdot + C_6H_5CO_2Ar + C_6H_5CO_2H$$

in which radicals produced in the reaction *induce decomposition* of further molecules of the peroxide. Induced decomposition is not usually observed for alkyl peroxides (e.g. di-t-butyl peroxide) but is characteristic of all acyl peroxide derivatives.

2.2. Photolysis

The simplest molecule is the hydrogen molecule and the simplest example of photolysis would be the fission of this molecule into two hydrogen atoms. In the ground state of the hydrogen molecule both electrons occupy the σ molecular orbital. When hydrogen absorbs light of a wavelength near 100 nm (in the vacuum ultraviolet) an electron is promoted to the σ^* molecular orbital, giving rise to an excited state in which both molecular orbitals are singly occupied. Since the σ^*-orbital is antibonding, the presence of the electron in this orbital roughly cancels the bonding power of the electron left in the σ-orbital, so that excited hydrogen molecules readily dissociate into atoms. The photochemical production of hydrogen atoms is not an important process, but the photolysis of halogen molecules, especially chlorine and bromine, is extremely important. Photolysis of both can be achieved by visible or near-ultraviolet light.

$$Cl_2 \xrightarrow{h\nu} 2Cl\cdot$$

$$Br_2 \xrightarrow{h\nu} 2Br\cdot$$

When we come to consider polyatomic molecules we have to consider not only excitation of σ-electrons but the excitation of electrons in π-orbitals and non-bonded electrons in atomic orbitals as well. It might seem that only σ → σ* transitions would be important in producing radicals but this is not so. The relative energies of orbitals are shown in figure 1.1. In hydrocarbons the σ → σ* transition occurs in the far ultraviolet and it is not much used as a source of hydrocarbon radicals. Available in the ordinary ultraviolet region are the π → π* transitions (e.g. the carbon–carbon double bonds in olefins); the n → σ* transitions (e.g. methyl iodide, involving non-bonding electrons of the iodine being promoted to the σ*-orbital) and the n → π* (the lowest of the electronic transitions, involving for example the non-bonded electrons on oxygen being promoted to the π*-orbital of a carbonyl group). n → π* Transitions are 'symmetry forbidden', but in practice the rigour of this rule is affected by vibrational motions which cause variation in the symmetry. The overall consequence is that n → π* transitions occur but the absorption is weak.

Both σ → σ* and n → σ* transitions can lead directly to radical formation. It is important to appreciate that the fragments of such a photochemical dissociation may not be in their ground state. For example methyl iodide is dissociated into methyl radicals and iodine atoms by light of 250 nm. However, light of this wavelength has more energy than that necessary to break the carbon–iodine bond. The result is that the iodine atom is released in an electronically excited state and the methyl radical is 'hot', i.e. possesses excess vibrational and translational energy.

$$\text{CH}_3\text{I} \xrightarrow[254 \text{ nm}]{h\nu} \text{CH}_3 \cdot{}^* + \text{I}(^2\text{P}_{1/2})$$

When an electron is excited from one orbital to another its spin remains unchanged. The initial absorption of light may be followed by the return of the electron to the ground state either with the emission of light (*fluorescence*) or through a descent of vibrational levels without emission (*internal conversion* and *vibrational quenching*). The important feature of both these processes is that they are fast, the initial absorption occurs in about 10^{-15} s, fluorescence in 10^{-5}–10^{-9} s, and non-radiative conversion in 10^{-9}–10^{-12} s. This means that the excited molecule will undergo few collisions before it returns to the ground state. However in many molecules a lower excited state of much longer life time is available; this is usually a triplet state (see p. 4). A change in relative electron spin is forbidden in the absence of perturbing effects. The usual perturbing factor is *spin–orbit coupling*, between the electron spin and orbital angular momentum. The process, '*intersystem crossing*', is quite fast (c. 10^{-6} s) but leads to the formation of the triplet which can have a comparatively long life (10^{-1}–10^{-5} s).

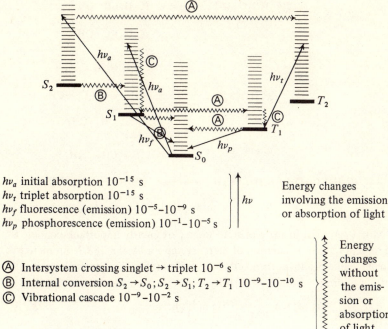

hv_a initial absorption 10^{-15} s
hv_t triplet absorption 10^{-15} s
hv_f fluorescence (emission) 10^{-5}–10^{-9} s
hv_p phosphorescence (emission) 10^{-1}–10^{-5} s

hv Energy changes involving the emission or absorption of light

(A) Intersystem crossing singlet → triplet 10^{-6} s
(B) Internal conversion $S_2 \to S_0$; $S_2 \to S_1$; $T_2 \to T_1$ 10^{-9}–10^{-10} s
(C) Vibrational cascade 10^{-9}–10^{-2} s

Energy changes without the emission or absorption of light

Figure 2.2. Modified Jablonski diagram showing the relationship between the ground state and various electronically excited states.

If it doesn't react chemically or transfer its energy to another molecule, a molecule in the triplet state will lose its energy either by emission (at a longer wavelength than the initial absorption - *phosphorescence*) or by internal conversion. For our present purposes it is the comparatively long life of the triplet state which is important. The triplet can undergo many collisions before it decays, and can take part in two types of reaction involving radicals.

The triplet state has many features characteristic of a radical and can be regarded as a biradical (see chapter 1, p. 4) and in some cases the triplet will take part in radical transfer reactions. For example the triplet of benzophenone will abstract hydrogen atoms from weak carbon–hydrogen bonds.

$$(C_6H_5)_2C{=}O \xrightarrow{hv} (C_6H_5)_2C{=}O^* \rightsquigarrow (C_6H_5)_2\dot{C}{-}\dot{O}$$

'singlet'   'triplet'

$$(C_6H_5)_2\dot{C}{-}\dot{O} + (CH_3)_2CHC_6H_5 \longrightarrow (C_6H_5)_2\dot{C}{-}OH + (CH_3)_2\dot{C}C_6H_5$$

Cumene

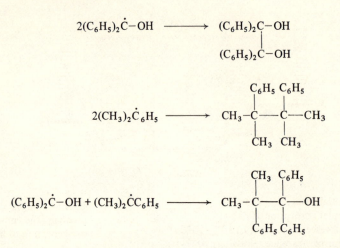

$$2(C_6H_5)_2\dot{C}-OH \longrightarrow \begin{array}{c} (C_6H_5)_2C-OH \\ | \\ (C_6H_5)_2C-OH \end{array}$$

$$2(CH_3)_2\dot{C}_6H_5 \longrightarrow CH_3-\overset{\overset{\displaystyle C_6H_5}{|}}{\underset{\underset{\displaystyle CH_3}{|}}{C}}-\overset{\overset{\displaystyle C_6H_5}{|}}{\underset{\underset{\displaystyle CH_3}{|}}{C}}-CH_3$$

$$(C_6H_5)_2\dot{C}-OH + (CH_3)_2\dot{C}C_6H_5 \longrightarrow CH_3-\overset{\overset{\displaystyle CH_3}{|}}{\underset{\underset{\displaystyle C_6H_5}{|}}{C}}-\overset{\overset{\displaystyle C_6H_5}{|}}{\underset{\underset{\displaystyle C_6H_5}{|}}{C}}-OH$$

Atoms can exist in a triplet state and a particularly important example is the mercury atom. Irradiation of mercury vapour by the 253.7 nm mercury resonance line produces the excited atom Hg $6(^3P_1)$. This excited atom will abstract hydrogen from molecular hydrogen or from alkanes.

$$Hg\ 6(^1S_0) \xrightarrow[253.7\ nm]{} Hg\ 6(^3P_1)$$

$$Hg\ 6(^3P_1) + H_2 \longrightarrow HgH + H\cdot$$

$$HgH \longrightarrow Hg\ 6(^1S_0) + H\cdot$$

HgH is an unstable intermediate which has only been detected recently. This reaction has been used quite extensively as a source of alkyl radicals, where mercury need only be present in catalytic amounts.

$$RH \xrightarrow[253.7\ nm]{Hg} R\cdot + H\cdot$$

The production of radicals by photo-initiated processes which do not involve simple photolysis includes photosensitization. This process starts by the absorption of light by one molecule which on collision passes this energy on to another molecule which then undergoes a chemical reaction. An important example again involves mercury atoms. Ethylene undergoes decomposition to acetylene and molecular hydrogen when irradiated in the presence of mercury vapour.

$$Hg\ 6(^1S_0) \rightarrow Hg\ 6(^3P_1)$$

$$Hg\ 6(^3P_1) + CH_2{=}CH_2 \rightarrow Hg\ 6(^1S_0) + (CH_2{=}CH_2)^*$$

$$(CH_2{=}CH_2)^* \rightarrow HC{\equiv}CH + H_2$$

$$(CH_2{=}CH_2)^* + CH_2{=}CH_2 \rightarrow 2CH_2{=}CH_2$$

The excited mercury atom transfers its energy to an ethylene molecule which then decomposes (recent work suggests two separate excited states of ethylene may be involved).

The final source of radicals involving electromagnetic radiation is radiolysis – that is the use of γ-rays or X-rays. There are three principal ways in which γ-rays interact with matter: there is the *photoelectric effect* in which the entire energy of the γ-photon is transferred to an atomic electron, with the subsequent ejection of the electron; there is the *Compton effect*, in which the γ-photon only gives up part of its energy to an electron and the incident photon is scattered and proceeds with an energy of the original quantum less the recoil energy of the electron; and there is *pair production*, in which the γ-photon is completely absorbed within the vicinity of an atomic nucleus and produces a positron-electron pair which after being slowed down recombines and emits a new γ-photon of much lower energy. In photochemistry one quantum will be entirely absorbed by one molecule to give, at least initially, one excited state. In radiation chemistry the initial quantum will bring about the excitation and ionization of many molecules. Excitation of molecules by slow electrons is not governed by selection rules, and states not formed in an optical primary process (e.g. triplets) may be formed. It is clear that a complex mixture of products is to be expected from radiolysis; for example with methane the following processes are postulated.

$$CH_4 \xrightarrow{\wedge\wedge\wedge} CH_4^+ + e^-$$

$$CH_3^+ + H \cdot + e^-$$

$$CH_2^+ + H_2 + e^-$$

$$CH_4^* \text{ (electronically excited)}$$

The major products from the radiolysis of methane are hydrogen and ethane, but the minor products include ethylene, propane, butanes, isopentane, acetylene and polymer. The importance of radiolysis is not the complex range of minor products but the formation of radicals which will then take part in chain reactions. For example the radiolysis of carbon tetrachloride solutions in cyclohexane has been used to study the reactions of trichloromethyl radicals.

$$2RH \xrightarrow{\wedge\wedge\wedge} 2R \cdot + H_2 \text{ (overall process)}$$

$$R \cdot + CCl_4 \longrightarrow RCl + CCl_3 \cdot$$

$$CCl_3 \cdot + RH \longrightarrow CCl_3 H + R \cdot$$

2.3. Oxidation

If an anion loses one electron it becomes a neutral radical. Processes of this kind are extremely important, particularly in living systems (see chapter 16) and will not be discussed in detail here. A single example is shown.

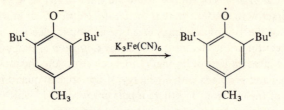

In addition to chemical oxidation, electrochemical oxidation is also important. The Kolbe synthesis using carboxylic acids will be described in chapter 13. An extension of this reaction is shown below.

$$RCO_2^- \xrightarrow[(-CO_2)]{-e} R\cdot \xrightarrow{Kolbe} \tfrac{1}{2}R_2$$

2.4. Reduction

Reductions are a less familiar source of radicals, probably because carbonium ions are less common as stable intermediates than carbanions. None the less, the addition of an electron to a carbonium ion results in the formation of a radical. A simple example is the reduction of the stable triphenylmethyl cation to the 'stable' free radical triphenylmethyl by vanadous chloride.

$$(C_6H_5)_3C^+ + V^{2+} \rightarrow (C_6H_5)_3C\cdot + V^{3+}$$

Just as electrochemical oxidation is possible at the anode of a cell, so electrochemical reduction is possible at the cathode.

3 Electron spin resonance spectroscopy

3.1. Introduction

Electron spin resonance spectroscopy is concerned with the study of species containing one or more unpaired electrons and hence can be applied to the study of radicals. The technique involves placing a specimen in a strong magnetic field and observing resonance effects in a radar circuit surrounding the specimen.

Detailed analysis of the ESR spectrum frequently makes it possible to deduce not only the gross chemical structure of a radical but also its conformation. In certain radicals, e.g. the phenoxy radical, the unpaired electron is delocalized throughout the radical. The electron density at any one position is referred to as the spin density and may be estimated by ESR spectroscopy. The sensitivity of the method enables radical concentrations of 10^{-8} M to be observed.

3.2. Principles of ESR spectroscopy

The principles underlying ESR spectroscopy closely resemble those of the much more familiar NMR spectroscopy. The electron with its spin has an associated magnetic moment just as does the proton. Consequently the electron will precess in an applied magnetic field with a precise precessional frequency and will undergo transitions between spin states if energy of the correct frequency is applied. The difference in energy of these spin states is given by the equation:

$$\Delta E = h\nu = g\mu_{B}B$$

where g is a dimensionless proportionality constant (often referred to as the g-factor), μ_B is the Bohr magneton (= $eh/4\pi m_e$, where e and m_e are the charge and mass of the electron respectively and h is Planck's constant) and B is the magnetic induction.* The frequency associated with a transition from

---

*See Note on units (pp. xv and xvi) for units used in this equation and throughout this chapter.

one energy level to the other is thus given by the equation:

$$\nu = \frac{g\mu_B B}{h}$$

There are thus two distinct ways in which a spectrometer could be designed to observe the ESR absorption: (*a*) the frequency could be altered using a fixed magnetic induction or (*b*) the magnetic induction could be varied using a fixed frequency.

In practice for the study of organic radicals a frequency of about 9000 MHz, provided by a microwave source, is employed while the magnetic induction is varied. At resonance some of the incident microwave radiation is absorbed: the value of *B* for resonance is about 0.33 T. In contrast to NMR spectrometers, ESR spectrometers are arranged to record the first derivative of the absorption curve rather than the absorption curve itself (figure 3.1).

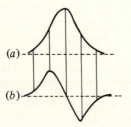

Figure 3.1. Curves used in the presentation of ESR spectra: (*a*) absorption curve; (*b*) first derivative of absorption curve.

This gives somewhat greater sensitivity and also better resolution. The area under the absorption curve is proportional to the number of spins in the sample. Integration of the first derivative to give the absorption curve followed by integration of this to obtain its area enables one to determine radical concentrations by comparison of this area with that due to a known concentration of radicals. Fremy's salt, $(KSO_3)_2 NO \cdot$, is frequently used as a standard.

Sufficiently high steady-state concentrations of radicals for ESR studies are achieved by the use of three techniques: (*a*) generation of radicals in a matrix at very low temperatures: (*b*) ultraviolet or X-ray irradiation of a solution of a radical precursor, e.g. a peroxide or an azo-compound, within the cavity of the ESR spectrometer; or (*c*) use of a flow system. In the flow technique two reactants which generate the radicals flow separately into a mixing chamber and then the mixture flows through the spectrometer cavity. By using rapid flow rates and placing the mixing chamber as close to the cavity as possible, radicals can be detected within 10^{-2} s of their formation.

3.3.　Characteristics of ESR spectra

Electron spin resonance spectra are characterized by three parameters: *g*-factors, hyperfine splitting constants, and line widths. A close study of these parameters enables much detailed structural information about the particular radical to be deduced.

3.3.1.　*g-factors*

In a magnetic field an unpaired electron in a radical possesses, in addition to its spin angular momentum, a small amount of extra orbital angular momentum. The interaction between these, called spin–orbit coupling, results in the electron having a slightly different effective magnetic moment from that of the free electron and consequently the condition for resonance is altered ($h\nu = g\mu_B B$ where *g* is dependent on the radical). Hence for a given frequency, radicals with different *g*-factors resonate at different field strengths. The difference in the *g*-factor for a radical and that for the free electron is analogous to the chemical shift in NMR spectroscopy. These differences are small, but they nevertheless can give valuable information about the structure of a radical.

Strictly speaking the spin–orbit coupling of the electron in a radical depends on the orientation of the radical with respect to the applied magnetic induction (the significance of this is discussed in chapter 16, p. 183). This is, however, unimportant for small radicals in solution where there is rapid tumbling of the radicals and consequently one obtains a time-averaged value for the *g*-factor.

3.3.2.　*Hyperfine splitting*

This is by far the most useful characteristic of ESR spectra both for elucidating the structure and also the shape of the radical under study. It arises from interaction between the unpaired electron and neighbouring magnetic nuclei (^1H, ^{13}C, ^{14}N, ^{17}O, etc.).

In a hydrogen atom the electron interacts with the proton which has a spin, I, equal to ½. The proton spin has two possible orientations ($m_I = \pm\frac{1}{2}$) which are parallel and antiparallel with the electron spin ($m_S = \pm\frac{1}{2}$) so that four energy levels result (see figure 3.2). The nuclear spin quantum number, m_I, may not change during an electronic transition so there are two allowed transitions which are indicated by arrows in figure 3.2. The ESR spectrum of the hydrogen atom therefore consists of two lines of equal intensity. The separation of the two lines is known as the *hyperfine coupling constant*, or *hyperfine splitting constant*: it is measured in millitesla (mT) using the SI system of nomenclature, or in gauss (1 mT = 10 gauss).

Interaction of the unpaired electron with three magnetically equivalent protons as in the methyl radical results in a 1 : 3 : 3 : 1 quartet (figure 3.3). Interaction with one and two magnetically equivalent protons will give rise to

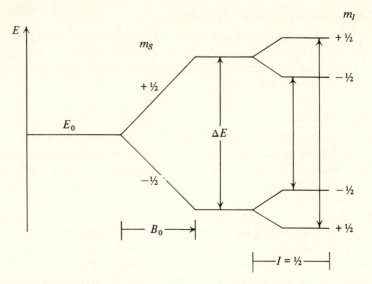

Figure 3.2. Splitting of electron energy levels in a magnetic field B_0 ($\Delta E = g\,\mu_\beta\,B_0 = h\nu$). Allowed transitions are indicated by arrows.

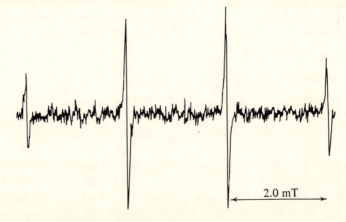

Figure 3.3. ESR spectrum of the $CH_3\cdot$ radical.

a 1 : 1 doublet and a 1 : 2 : 1 triplet respectively, and in general interaction with n equivalent protons gives $n + 1$ lines whose relative intensities are given by the coefficients of the binomial expansion. Interaction of the unpaired electron with a nucleus of spin I gives $(2I + 1)$ lines. Thus the ESR spectrum of a radical centred on nitrogen ($I = 1$) consists of a 1 : 1 : 1 triplet (figure 3.4).

ESR spectra are in general much more complex than NMR spectra because coupling with β-protons is frequently as great as if not greater than that with

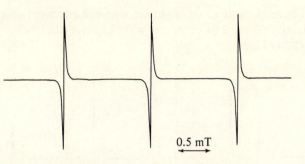

Figure 3.4. ESR spectrum of Fremy's salt, $(KSO_3)_2 NO \cdot$

α-protons. Thus the spectrum of the ethyl radical consists of twelve lines (figure 3.5). The spin of the unpaired electron is split into a quartet by the β-protons of the methyl group and each of these lines is split into three lines by the protons of the methylene group. The coupling constants with the α- and β-protons are 2.24 and 2.69 mT respectively.

Careful analysis of the hyperfine splitting enables assignments to be made as to radical structure. The spectrum of the radical resulting from oxidation

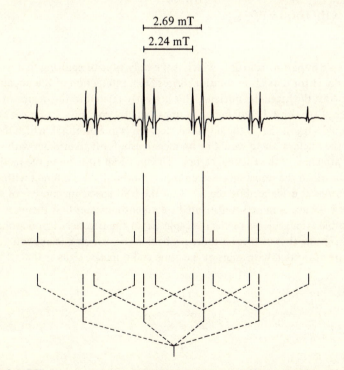

Figure 3.5. ESR spectrum of the ethyl radical with the stick diagram and interpretation of the spectrum.

of ethanol by hydroxyl radicals was a quartet, each component of which was split into a doublet (figure 3.6), consistent with the $CH_3\dot{C}HOH$ radical but not the $\dot{C}H_2CH_2OH$ radical.

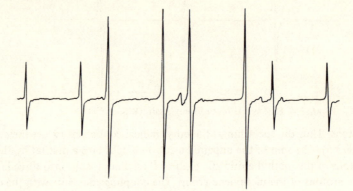

Figure 3.6. ESR spectrum of the $CH_3\dot{C}HOH$ radical.

$$CH_3CH_2OH + HO\cdot \begin{array}{c} \nearrow CH_3\dot{C}HOH \\ \searrow\!\!\!\times \dot{C}H_2CH_2OH \end{array}$$

Isotropic hyperfine splitting, which is the only type of splitting that need be considered for radicals in solution, arises from interaction of the unpaired electron with the magnetic nucleus. This is clearly related to the degree of s-character of the orbital containing the unpaired electron. First-order considerations would suggest that if the unpaired electron is in a p-orbital, which has a node at the nucleus, there would be no mechanism for it to interact with protons attached to the trigonal carbon. The unpaired electron in the methyl radical, in which the unpaired electron is in a p-orbital, does interact with the three protons as evidenced by the fact that the ESR spectrum consists of a $1:3:3:1$ triplet. A more sophisticated treatment indicates that there are two possible arrangements of electron spin about the trigonal carbon atom (figure 3.7): the first of these is more probable. The carbon nucleus and proton are associated with spins of the same and opposite signs as that of the

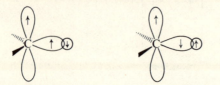

Figure 3.7. Orbital diagram showing the two possible spin polarizations in the $C-H$ σ bond.

unpaired electron. There is thus a net unpaired negative spin density in the *s*-orbital of the proton which gives rise to hyperfine splitting with a negative sign. The mechanism of this interaction is described as *spin polarization.* In contrast the hyperfine splitting constant is positive and of greater magnitude if the unpaired electron is in an *s*-containing orbital (see p. 38).

The unpaired electron is also coupled to protons attached to β-carbon atoms in the radicals as a result of overlap of the $2p_z$ orbital containing this electron and the β-C−H bonding orbital. This gives a spin population in the 1*s* orbital of the hydrogen atom. In valence-bond terms there is a contribution to the resonance hybrid of the canonical structure (2). The spin of the unpaired electron in the $2p_z$ orbital is paired with the opposite spin in the C−H bond. The β-proton is thus associated with spin density of the same sign as that in the $2p_z$ orbital and can be said to have positive spin density thus giving rise to hyperfine splitting of positive sign.

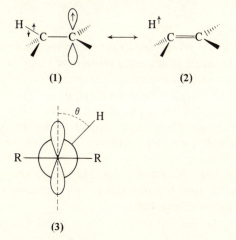

There is a clear requirement for the occurrence of this interaction that the geometry of the system should permit overlap of the appropriate orbitals. Stereochemically the overlap is greatest when the dihedral angle, θ, between the $2p_z$ orbital and the β-C−H bond (cf. **3**) is 0° and least when this angle is 90°, i.e. when the orbitals are perpendicular to each other. The magnitude of the splitting due to this type of interaction is proportional to $\cos^2 \theta$. The value of β-proton coupling constants and their dependence on temperature can give information about preferred conformations of radicals and of the barriers to rotation about $C_\alpha−C_\beta$ bonds (p. 39). (That the protons in β-C−H bonds interact with an unpaired electron and that the magnitude of this interaction is dependent on the angle θ, provides perhaps the best experimental evidence for hyperconjugation.)

Hyperfine splitting with γ-protons can also be observed on occasion.

Small long-range splittings are observed in certain bicyclic radicals of fixed geometry. These are greatest when the bonds between the radical centre and the proton are in a planar W-shaped arrangement. An example of this phenomenon is seen in the semidione radical (4) in which the coupling constant with the *anti*-hydrogen at C-7 is much greater than that with the *syn*-hydrogen. (Similar long-range interactions are also encountered in NMR spectroscopy.)

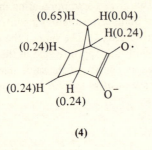

(4)

Hyperfine splitting constants in substituted alkyl radicals are also suscept-ible to the ability of the substituents to delocalize the unpaired spin: both electron-donating and electron-withdrawing substituents can delocalize un-paired spin. Thus the proton splitting constants in radicals of the type $\dot{C}H_2X$ and $\dot{C}HX_2$ (where X = OR, CN, COR, CO_2R, Me etc.) are lower than that of the methyl radical provided that the geometries of the radicals are similar. (Splitting constants vary somewhat with the geometry of the radical, see p. 38.)

The magnitude of the coupling constant of a proton (a_i) is directly propor-tional to the spin density on the atom to which that proton is attached (ρ_i). This is represented by the McConnell equation:

$$a_i = \rho_i Q$$

where Q is a proportionality factor. This equation is of great value in calculat-ing spin densities in delocalized radicals. Thus measurement of the three hyperfine splitting constants in the ESR spectrum of the phenoxy radical (figure 3.8) enables the relative electron densities at the *ortho*-, *meta*-, and *para*-positions to be calculated. If Q is known then the actual spin densities ρ_o, ρ_m and ρ_p at these positions can be determined. Theoretical considera-tions, which are outside the scope of this book, indicate that the spin density at the *m*-position is in fact negative. The spin density at oxygen, ρ^o, can be obtained to a first approximation by difference:

$$\rho^o = 1 - (2\rho_o + 2\rho_m + \rho_p)$$

In this way it can be estimated that the spin density on oxygen in a phenoxy radical is about 0.2 while at the *ortho*- and *para*-carbon atoms the spin

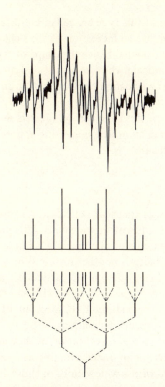

Figure 3.8. ESR spectrum of the phenoxy radical with the stick diagram and interpretation of the spectrum.

densities are approximately 0.2 and 0.4 respectively. The spin density at oxygen can only be obtained directly using the ^{17}O-labelled phenoxy radical (^{17}O possesses spin and hence can interact with the unpaired electron). Similarly the spin density at C-1 can only be obtained by use of a ^{13}C-labelled phenoxy radical.

Measurement of hyperfine splitting constants can also be used to estimate the spatial configuration of phenyl substituents in, for example, the 2,4,6-triphenylphenoxy radical in which the unpaired electron is delocalized onto the phenyl substituents. Theoretical calculations are made with the phenyl substituents at different angles to the phenoxy ring to obtain the best fit between the experimental and theoretical values. For the 2,4,6-triphenylphenoxy radical this obtains when the 2- and 6-phenyl substituents are inclined at an angle of 46° to the phenoxy ring and the 4-phenyl substituent is coplanar with the phenoxy ring.

Coupling with ^{13}C is very valuable in obtaining information about radical shapes: ^{13}C splitting constants are very much larger if the unpaired electron is in an s-containing orbital (such radicals are termed σ-radicals) than if the unpaired electron is in a *p*-orbital (radicals of this type are called π-radicals). [13]C splitting constants can thus be used to give an indication of the *s*-character of the orbital containing the unpaired electron and hence the shape of the radical. This technique has been used to indicate whether substituted methyl radicals are planar or slightly pyramidal (the unpaired electron is in an sp^3-orbital in a radical with tetrahedral geometry) (see p. 37).

The very large ^{13}C coupling constants for vinyl and formyl radicals show that in both cases the unpaired electron is in an orbital possessing appreciable *s*-character and hence that these are σ-radicals.

3.4.   Spin trapping

The direct detection by ESR spectroscopy of transient radicals as reaction intermediates is restricted in the main to studies of flow systems. This procedure requires the use of relatively large amounts of material and is limited in scope. The problem then arises as to how radicals in other reactions may be detected. This problem has been solved by carrying out the particular reaction in the presence of a suitable radical scavenger which can react with any radical intermediate to form a relatively long-lived radical, the ESR spectrum of which can be obtained without recourse to special techniques.

The two most commonly used types of radical scavenger or, as they are frequently referred to 'spin trap' are nitroso-compounds and nitrones (reactions *1* and *2*). Both of these react rapidly with a wide variety of short-lived

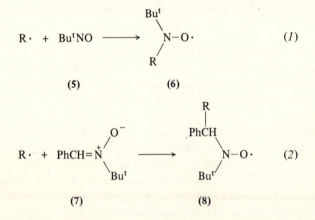

radicals to give relatively long-lived nitroxides. The ESR spectra of these resultant nitroxides are characterized by a 1 : 1 : 1 splitting due to interaction of the unpaired electron with the ^{14}N nucleus which has a spin quantum number, *I*, of 1. This triplet is further split as a result of coupling with magnetic nuclei on the α- and β-carbon atoms. t-Nitrosobutane **(5)** and *N*-t-butyl α-phenyl-nitrone **(7)** are very frequently used since the protons of the t-butyl groups only interact to a very minimal extent with the unpaired spin. In general nitrones are less attractive as spin traps than nitroso-compounds since the radical fragment is further from the radical centre in (8) than in (6) and thus contributes less to the splitting pattern. The magnitude of the nitrogen coupling constant is also sensitive to the type of radical trapped and one can for example in this way distinguish between a carbon radical and an oxygen radical.

An example of the use of t-nitrosobutane as a spin trap is seen in figure 3.9

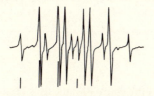

Figure 3.9. The ESR spectrum of t-butyl methyl nitroxide.

which records the ESR spectrum of t-butyl methyl nitroxide which is formed by trapping methyl radicals, produced in the decomposition of diacetyl peroxide, with t-nitrosobutane. The spectrum consists of three 1 : 3 : 3 : 1 quartets as a result of interaction of the unpaired spin with the nitrogen nucleus and the three protons of the methyl group.

Spin traps have also been used quantitatively to determine rate constants for radical processes. The technique involves competition between bimolecular spin-trapping and a reaction of known rate such as the unimolecular fragmentation of the radical undergoing trapping.

4 Chemically induced dynamic nuclear polarization (CIDNP)

4.1. The phenomenon

Direct application of NMR spectroscopy to the study of radicals is limited to concentrated solutions of stable forms. However, the presence of transient radicals can be detected, and information about their structure and mechanism of reaction can be deduced, from NMR spectra recorded during the progress of certain chemical reactions. Protons in a molecule are normally partitioned among their spin states according to the Boltzmann distribution. For example, one proton has two available spin states, and because the energy difference is very small, these two levels are almost equally populated. However, in the products of some radical reactions this Boltzmann distribution may be significantly perturbed so that there are excess protons in the upper or lower spin states. When this occurs the magnetic nuclei spontaneously emit or absorb radiation until they have relaxed into the equilibrium situation. The NMR spectrum of the product recorded while the magnetic nuclei are reverting to the equilibrium distribution will show greatly increased intensity of absorption for some lines, and negative peaks indicating emission of radio-frequency radiation for other lines. These two possibilities are given the symbols A for enhanced absorption and E for emission.

An excellent example of the nuclear polarization which is induced in radical reactions is provided by the decomposition of acetyltrichloroacetyl peroxide in carbon tetrachloride containing iodine. The peroxide decomposes at $50°C$ to give methyl and trichloromethyl radicals which can combine within the solvent cage to give 1,1,1-trichloroethane or escape and react with iodine.

$$CH_3COOCCCl_3 \xrightarrow{\;50\,°C\;} [CH_3\cdot\;\cdot CCl_3]_{cage} \xrightarrow[\;I_2\;]{\text{diffusion}} CH_3I + CCl_3I$$
$$\underset{O\quad O}{\overset{\parallel\quad\parallel}{}}$$
$$\downarrow$$
$$CH_3CCl_3$$

The ^{1}H NMR spectrum recorded at $50°C$ five minutes after the start of the reaction is shown in the upper trace of figure 4.1. The signal for the protons

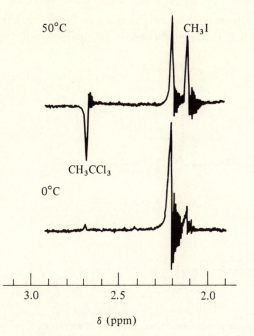

Figure 4.1. Transient NMR spectrum during the thermal decomposition of acetyltrichloroacetyl peroxide in carbon tetrachloride containing iodine. From H. R. Ward,
Accounts of Chemical Research, 1972, **5**, 18; reproduced by permission of *Accounts
of Chemical Research*.

of 1,1,1-trichloroethane appears at $\delta = 2.7$ as a negative peak, indicating
emission from this product, and the scavenged product methyl iodide at
$\delta = 2.1$ shows enhanced absorption. The signal at $\delta = 2.2$ is from the parent
peroxide. The lower trace shows the spectrum of the same mixture after the
reaction had been halted by cooling to $0°C$. The signals for both products are
showing normal absorption, and their small size in comparison with the upper
trace indicates the large intensity enhancement which arises from chemically
induced nuclear polarization.

The protons of 1,1,1-trichloroethane show *net* emission and those of
methyl iodide show *net* absorption, but sometimes both emission and absorption can occur in a single multiplet formed by equivalent protons. This is
observed in the ethyl chloride formed in the decomposition of dipropanoyl
peroxide in hexachloroacetone.

$$\underset{\substack{\| \ \| \\ O \ \ O}}{EtCOOCEt} \xrightarrow{\Delta} [Et\cdot \ \cdot Et]_{cage} \xrightarrow{CCl_3COCCl_3} CH_3CH_2Cl$$

$$\downarrow \text{combination}$$

The transient NMR spectrum of the ethyl chloride run during the decomposition is shown in figure 4.2. The low-field lines of the methylene quartet at

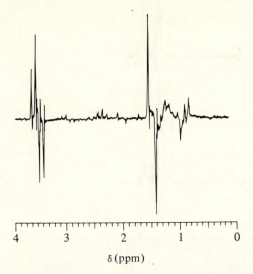

δ (ppm)

Figure 4.2. NMR spectrum of the decomposition products of dipropanoyl peroxide in hexachloroacetone. From R. Kaptein and L. J. Oosterhoff, *Chemical Physics Letters*, 1968, **2**, 261; reproduced by permission of *Chemical Physics Letters*.

δ = 3.6 show enhanced absorption, and the high-field lines show emission which almost exactly counterbalances it. Similarly, the low-field line of the methyl triplet at δ = 1.5 shows enhanced absorption, the high-field line shows emission, while the central line of the triplet shows virtually no enhancement and is not observable on the scale shown. The CIDNP spectrum of a given product can show the *net effect* which is either enhanced absorption *A* or emission *E*, or it can show the *multiplet effect* in which there is no net enhancement, but either first enhanced absorption and then emission with increasing field, *AE* (as in figure 4.2) or emission followed by enhanced absorption, *EA*. In practice, many CIDNP spectra show a combination of the two types of behaviour.

Nuclear polarization giving rise to NMR spectra of this kind is confined to the products of radical reactions. Anionic and cationic intermediates do not lead to polarized products, nor do concerted reactions. The observation of a CIDNP spectrum is therefore positive evidence that radical intermediates are present. It is possible that a small fraction of the product molecules could be formed by a radical route, thus giving a CIDNP spectrum, while the bulk of the product was formed by an ionic or concerted pathway. CIDNP spectra are not observable on the products of all radical reactions because the lifetime of the nuclear polarization may be too short. Therefore the absence of a CIDNP spectrum is not compelling evidence that radicals are absent.

4.2. The radical pair theory

According to the established theory of CIDNP spectra, the nuclear polarization owes its origin to the pairing of radicals formed in the initiation process,
or by radicals encountering one another in the solution. When a molecule
$R^1 - R^2$ decomposes homolytically in solution a pair of radicals is formed
within a cage of solvent molecules. The radicals may react within the cage by
combination and/or disproportionation to give *geminate* products, or they
may diffuse out into the bulk of the medium and react with some other
molecule to give *scavenged* products.

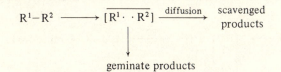

Before the radicals separate the electrons are still weakly coupled and, therefore, the radical pair will be in either a singlet state or a triplet state depending
on its mode of formation. Thermal decomposition of peroxides or azo-
compounds gives radicals with electrons having antiparallel spins so that the
radical pair will be in a singlet state. On the other hand, photochemical generation of radicals frequently, though not invariably, leads to triplet radical pairs.

The first assumption of the radical pair theory is that singlet radical pairs
react more readily to give geminate products than do triplet radical pairs.
This follows from the fact that a pair of electrons with antiparallel spins can
unite at once to form a bond, whereas in the triplet case, where the electrons
have parallel spins, spin inversion must take place before bond formation can
occur.

$$\overline{[R^1\uparrow \downarrow R^2]}^S \xrightarrow{\text{fast}} R^1 - R^2$$

$$\Big\uparrow \begin{array}{l} \text{spin} \\ \text{inversion} \end{array}$$

$$\overline{[R^1\uparrow \uparrow R^2]}^T \xrightarrow{\quad\times\quad} R^1 - R^2$$

We consider next a radical pair which is initially in a singlet state. The electrons in this pair experience a local magnetic field which is the sum of the
applied field B_0 and internal fields arising from electron orbital motion and
nearby nuclear spins. As time goes by the radical pair will tend to develop
triplet character and this tendency will be reinforced or opposed by the local
field. The interaction of the electron with neighbouring orbital motion is
given in terms of the g-factor of the radical, so the rate of crossing from
singlet to triplet will depend on the difference in g-factors of the two radicals.
The influence of nearby magnetic nuclei depends on their hyperfine coupling

constants with the electron (a) and their spin quantum numbers (m). The rate of crossing depends therefore on the following two terms:

$$(g_1 - g_2)B_0 + (\Sigma a_1 m_1 - \Sigma a_2 m_2)$$

Here g_1 and g_2, the g-factors of the two radicals, are usually known from conventional ESR experiments. The second term arises from the sum of the contributions of all the magnetic nuclei on R^1 and all the magnetic nuclei on R^2.

The simple case in which the first of the radicals in the singlet pair contains one type of proton and the second has no magnetic nuclei coupled to the electron will now be considered:

$$HM-N \longrightarrow [HM\uparrow \ \downarrow N]^S \xrightarrow{\ RX\ } HM-X \ + \text{ other products}$$
$$\underset{g_1 \quad\quad g_2}{}$$
$$\downarrow$$
$$HM-N$$

If the two radicals have differing g-values such that $g_1 < g_2$ then the first term in the above equation is negative and as time progresses the radical pair will develop more triplet character. In a proportion of the radicals HM · the protons will be in the upper spin state, i.e. $m_1 = -\frac{1}{2}$ and if they are coupled to the electron by a negative coupling constant, i.e. a_1 is negative as in methyl, (see p. 24) the second term in the equation will be positive and the nuclear spin term will retard the rate of crossing to the triplet. Hence radical pairs in which the protons are in the upper state will retain singlet character, and because singlet radical pairs react more readily to give geminate products, it follows that the geminate products will be formed with excess population of protons in the upper state. The equilibrium spin population will be restored by emission from the geminate product as protons return to the lower state, and this will lead to a negative peak in the NMR spectrum. The geminate product CH_3CCl_3 from acetyltrichloroacetyl peroxide decomposition shows exactly this behaviour (see figure 4.1).

On the other hand, for the proportion of the radicals HM · with protons in the lower state ($m_1 = +\frac{1}{2}$) the nuclear spin contribution is negative, and the rate of crossing to the triplet will be augmented. Since cage combination is unlikely for the triplet pairs there will be a higher probability of these radicals diffusing out into the solution and forming scavenged products. Hence the scavenged products will be rich in protons in the lower spin state and they will show enhanced absorption as the equilibrium distribution is restored. The enhanced absorption shown by the scavenged product from acetyltrichloroacetyl peroxide (see figure 4.1) illustrates this type of behaviour.

If the original radical pair had been formed in a triplet state, but all the other parameters were the same, then the coupling of radical HM · with protons in the upper ($-\frac{1}{2}$) state would tend to retard the rate of crossing to the singlet. Hence the scavenged products would be rich in the upper state, and it is clear that the polarization of both geminate and scavenged products would be opposite to that of the singlet radical pair. It also follows from this analysis that the direction of the polarization depends on the relative values of the g-factors of the two radicals and on the sign of the electron–nuclear coupling constants of magnetic nuclei in the radicals.

The results of radical pair theory can be summarized in two rules known as *Kaptein's rules*, which enable the sign of the CIDNP effect to be predicted. The rules are based on a number of approximations and can break down at low magnetic fields and under certain other conditions. Considering a pair of radicals with g-factors g_a and g_b containing nuclei i and j respectively with coupling constants a_i and a_j, having nuclear–nuclear coupling constants J_{ij}, the net effect, Γ_{ne} is given by:

$$\Gamma_{ne} = \mu\epsilon\Delta ga_i \quad \begin{cases} +\text{ve } \Gamma_{ne} \text{ indicates } A \\ -\text{ve } \Gamma_{ne} \text{ indicates } E \end{cases}$$

The signs of the parameters in this expression are as follows:

$$\mu = \begin{cases} + \text{ for triplet radical pair or for encounter of freely diffusing radicals} \\ - \text{ for singlet radical pair} \end{cases}$$

$$\epsilon = \begin{cases} + \text{ for geminate products} \\ - \text{ for scavenged products} \end{cases}$$

The second rule gives the direction of the multiplet effect, Γ_{me}:

$$\Gamma_{me} = \mu\epsilon a_i a_j J_{ij} \sigma_{ij} \quad \begin{cases} + \text{ ve } \Gamma_{me} \text{ indicates } EA \\ - \text{ ve } \Gamma_{me} \text{ indicates } AE \end{cases}$$

The symbols here have the same significance as before and σ_{ij} has the following meaning:

$$\sigma_{ij} = \begin{cases} + \text{ if nuclei } i \text{ and } j \text{ belonged to the same radical} \\ - \text{ if nuclei } i \text{ and } j \text{ belonged to different radicals} \end{cases}$$

Besides giving the simple diagnosis that the reaction involves radical intermediates, the CIDNP spectrum can be used to give information about the structure of the radical pairs involved. A judicious use of Kaptein's rules can enable data on the g-values of the radicals, the spin multiplicity of the radical pair, and the signs of the hyperfine coupling constants to be obtained or checked. For example, the transient NMR spectrum obtained during the thermal decomposition of diacetyl peroxide in CCl_4 showed E for the methoxy protons of the product methyl acetate, and A for the product methyl chloride. The following mechanism of decomposition may be proposed. The radical pair will be singlet since it is formed in a thermal process, hence μ is negative. Methyl acetate is formed by geminate combination of the methyl and acetoxy radicals, hence ϵ is positive. g-values of radicals tend to increase

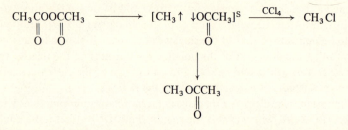

with the electronegativity of the radical, so it is expected that $g_{CH_3} - g_{AcO}$ will be negative. Hyperfine coupling constants of α-protons are usually negative and those of β-protons are usually positive. Thus application of the first rule for the net effect in the methoxy protons of methyl acetate shows

Γ_{ne} to be negative, indicating E.

Similarly, for the scavenged product CH_3I, ϵ changes sign but all the other parameters remain the same, and therefore A is predicted. Thus, Kaptein's rules support the validity of the proposed mechanism for diacetyl peroxide decomposition and confirm the multiplicity of the radical pair, the sign of Δg and of the hyperfine coupling constant of the methyl radical.

The thermal decomposition of dipropanoyl peroxide takes a slightly different course. The main product in hexachloroacetone solution is ethyl chloride which, as we have seen, shows no net effect but a multiplet effect in the sense AE (see figure 4.2). The polarization would appear to arise in this case from a pair of ethyl radicals in a singlet state (see above). Since both components of the radical pair are the same, Δg is zero and Γ_{ne} is therefore zero and no net effect is predicted. For the multiplet effect μ is negative for a singlet pair, ϵ is negative for the scavenged product, a_i is negative for α-protons, a_j is positive for β-protons, J_{ij} is positive and σ_{ij} is positive since both nuclei belonged to the same radical. The sign of the multiplet effect (Γ_{me}) for both methyl and methylene protons is given by

$(-)(-)(-)(+)(+)(+)$, which is negative, indicating AE.

The good agreement between these predictions and experiment is excellent evidence that the singlet ethyl radical pair is the key intermediate in the thermal decomposition of dipropanoyl peroxide.

CIDNP spectra have been extensively used in the study of peroxide decompositions, organolithium–alkyl halide reactions, ketone photolyses and in the investigation of radical pair rearrangements such as the Stevens rearrangement (see p. 168). The method is particularly effective when used in conjunction with ESR. Repeated scanning of the CIDNP spectrum during the course of a radical reaction can yield information about the history of radical pairs and the kinetics of the reactions involved.

5   Stereochemistry of free radicals

5.1.   Introduction

Alkyl radicals can be envisaged as having a planar structure, in which case the radical centre is sp^2-hybridized and the unpaired electron is in a p-orbital thus resembling the corresponding carbonium ion. Alternatively the radical may be pyramidal with an sp^3-hybridized radical centre. In this latter case there could be rapid inversion of the two mirror-image configurations.

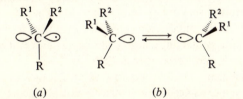

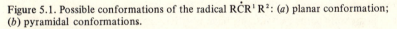

(a)   (b)

Figure 5.1. Possible conformations of the radical $R\dot{C}R^1R^2$: (a) planar conformation; (b) pyramidal conformations.

The shape of a radical may be deduced by use of either physical or chemical methods. Physical methods, which in the main involve ESR studies, are undoubtedly more satisfactory than chemical methods since one is dealing with the ground state of the radical. Chemical methods generally involve the generation of a radical at an asymmetric carbon atom and observation of whether or not racemization takes place. Retention (or inversion) of configuration would argue strongly in favour of a pyramidal radical. Racemization, however, is consistent with either a planar or a rapidly inverting pyramidal radical in which inversion proceeds more rapidly than reaction.

5.2.   Alkyl radicals

Pertinent information about the shape of alkyl radicals can be obtained from the ^{13}C splittings in their ESR spectra. ^{13}C splitting is a particularly sensitive measure of the degree of non-planarity of this type of radical, increasing as

Table 5.1. *Hyperfine splitting constants of methyl and fluorinated methyl radicals*

Radical	Coupling constants (mT)	
	$a(H)$	$a(^{13}C)$
$\cdot CH_3$	2.30	3.85
$\cdot CH_2F$	2.11	5.48
$\cdot CHF_2$	2.22	14.88
$\cdot CF_3$	–	27.16

the *s*-character of the orbital carrying the unpaired electron increases. Table 5.1 shows the H and ^{13}C coupling constants for the methyl radical and the fluorinated methyl radicals. The ^{13}C values steadily increase with the degree of fluorine substitution: that for the trifluoromethyl radical is approximately one quarter of the calculated ^{13}C splitting for an electron in a pure 2*s*-orbital. This is thus consistent with the unpaired electron in this radical being in an sp^3-orbital, i.e. with the trifluoromethyl radical being pyramidal. The ^{13}C splitting for the methyl radical is very much smaller. The low value can be considered to arise because the unpaired electron is in an orbital containing some *s*-character or as a result of spin polarization (see p. 24). The ^{13}C splittings for the $\dot{C}H_2F$ and $\dot{C}HF_2$ radicals are consistent with their having the unpaired electron in an orbital containing a degree of *s*-character, i.e. there is some bending in these radicals, more especially in $\dot{C}HF_2$.

The proton splitting of a planar radical is negative as a result of spin polarization (see p. 25), which induces a negative spin density at the trigonal carbon atom. Pyramidal radicals, in which the unpaired electron is in an orbital possessing some *s*-character, have positive proton splittings since there is a positive spin density associated with the radical centre. One can thus envisage the proton splittings of radicals numerically decreasing in value and then increasing as the degree of bending increases, since it is not possible to measure the signs of splittings in ESR spectra. On this basis it seems possible that the proton splitting of $\dot{C}H_2F$ is negative and that of $\dot{C}HF_2$ is positive when the proton splittings are considered along with the ^{13}C splittings (table 5.1). Infrared and ultraviolet studies on the methyl radical likewise support the contention that it adopts a planar conformation.

Groups other than fluorine which possess electron pairs also induce bending at the radical centre. Thus the $\dot{C}H_2OH$ and $\dot{C}Me_2OH$ radicals are non-planar.

Contrary to expectations the t-butyl radical, unlike the t-butyl cation, appears to be non-planar. Very bulky t-alkyl radicals such as $Bu^t_3C\cdot$ are, however, planar to minimize steric interaction amongst the bulky alkyl groups.

5.3. Cycloalkyl radicals

ESR studies indicate that the radical centre is essentially planar in cyclohexyl, cyclopentyl and cyclobutyl radicals. In contrast the cyclopropyl radical is distinctly bent as indicated by the α-proton splitting constant of 0.65 mT. This is supported by a number of chemical studies on optically active cyclopropyl compounds which indicate that cyclopropyl radicals react with partial retention of configuration at the radical centre.

5.4. β-Substituted alkyl radicals

Two extreme conformations can be envisaged for β-substituted alkyl radicals $XCH_2\dot{C}R_2$, namely the eclipsed conformation (1) and the staggered conformation (2). (In the eclipsed conformation the substituent X on the β-carbon atom is eclipsed with the half-filled p-orbital whilst in the staggered conformation the substituent X lies in the node of the half-filled p-orbital.) The dihedral

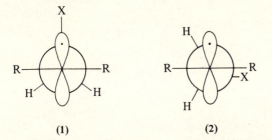

(1)   (2)

angles, θ, between the β-C—H bonds and the orbital containing the unpaired electron are 60° and 30° respectively in these conformations. This compares with an average dihedral angle of 45° for a freely rotating radical. The coupling constant for β-protons is related to the dihedral angle by the equation:

$$a_{\beta-H} = A + B \cos^2 \theta$$

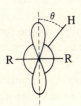

where A and B are constants and $B \gg A$. The magnitude of the β-C—H coupling constant for a radical in the eclipsed conformation (1) is thus greater than for a freely rotating radical which in turn is greater than for a radical in the staggered conformation (2). At a sufficiently high temperature, rotation about C—C bonds becomes increasingly free and consequently the coupling constant will approach that for a freely rotating radical, i.e. the coupling constant for a radical in the eclipsed conformation (1) will increase with increasing temperature while that for a radical in the staggered conformation (2) will decrease (see figure 5.2). ESR studies of this type have shown that the n-propyl radical

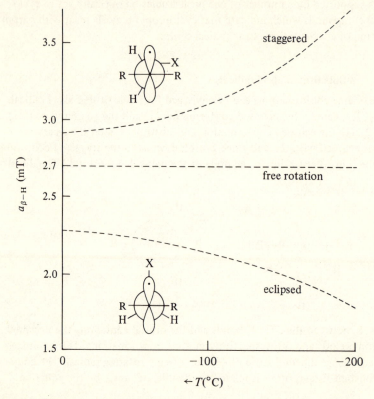

Figure 5.2. Temperature dependence of β-proton splitting constants, $a_{\beta-H}$, for β-substituted alkyl radicals R_2CCH_2X.

prefers the conformation (2) (R = H, X = Me) but the t-amyl radical favours the eclipsed conformation (1) (R = X = Me): the change in conformation is due to steric factors.

The barriers to rotation about $C_\alpha-C_\beta$ bonds can be obtained from the variation in the β-C—H coupling constant with temperature and also from the temperature dependence of line shapes.

ESR studies indicate that radicals of the type $R_3MCH_2CH_2\cdot$ (M = Si, Ge or Sn) invariably exist in the eclipsed conformation. The barrier to rotation about the $C_\alpha-C_\beta$ bond is much greater than that in alkyl radicals. This conformational preference is attributed to delocalization of the unpaired spin onto the $C_\beta-M$ bond through hyperconjugation and to $p-d$ homoconjugation.

	(1) Eclipsed conformation	Free rotation	**(2)** Staggered conformation
θ	$60°$	$45°$	$30°$
$\cos^2 \theta$	0.25	0.5	0.75
$a_{\beta - H}$	$A + \tfrac{1}{4}B$	$A + \tfrac{1}{2}B$	$A + \tfrac{3}{4}B$

A similar preference for the eclipsed conformation occurs with β-thioalkyl-ethyl radicals, $RSCH_2 \dot{C}H_2$.

The structure of β-haloalkyl radicals is of particular interest in view of the strong preference for *trans*-addition of hydrogen bromide and bromine to alkenes (see p. 104); these results have frequently been interpreted in terms of a 'bridged radical' analogous to a bromonium ion. Molecular orbital calculations indicate that whereas a non-classical or delocalized structure is possible for halonium ions it is most improbable for the corresponding radicals. ESR results indicate that β-chloroethyl radicals undoubtedly adopt an eclipsed conformation (**3**) in which the chlorine atom is bent back towards the orbital containing the unpaired electron: the chlorine atom is not symmetrically placed between C_α and C_β (see **4**), nor are the α and β protons equivalent. The radical is thus not bridged in the same sense as the analogous cation: it is, however, bridged in a broader sense due to the homoconjugative and hyper-conjugative interactions between the C—Cl bond and the unpaired electron. The $Me_2 \dot{C}CH_2 Br$ radical in contrast adopts the staggered conformation (**5**) in which the radical centre is definitely non-planar. The rotational barrier in this radical explains the observed 'retention' of stereochemistry in reactions involving it without recourse to invoking a bridged species. The difference in struc-

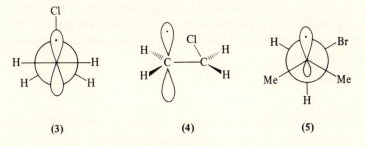

ture of the β-bromo- and β-chloroalkyl radicals is a consequence of the larger size of the bromine. Fluorine-containing radicals have a much greater degree of rotational freedom.

5.5. Allylic radicals

Allylic radicals show configurational stability. Thus the ESR spectra of the *cis*- and *trans*-1-methylallyl radicals (**6** and **7**) are quite distinct and there is no evidence of any isomerization of the radicals over a temperature range −130°C to 0°C. The acetonyl radical (**8**) and other 2-oxoalkyl radicals also have a high barrier to rotation about the C−CO bond.

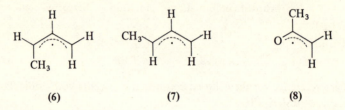

(6)   (7)   (8)

5.6. Vinylic radicals

Vinylic radicals could adopt either a bent or linear structure (**9** or **10**) according as to whether the carbon atom carrying the unpaired electron is sp^2- or sp-hybridized, i.e. the unpaired electron would be in an sp^2- or p-orbital respectively. The large ^{13}C splitting constant for the vinyl radical indicates that the

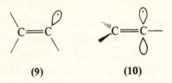

(9)   (10)

unpaired electron is in an s-containing orbital and hence favours the bent structure (**9**). Detailed analysis of the proton coupling constants in the ESR spectra of vinyl radicals is consistent with a bent structure with rapid inversion of its configuration even at −180°C (**11** and **12**). The ESR spectrum of the 1-methylvinyl radical showed two distinct β-proton splittings indicative of two non-equivalent protons and hence of a non-linear structure (**13**) which is configurationally stable at the temperature of the study. The greater size of the methyl group with respect to hydrogen is credited with causing this decreased rate of inversion.

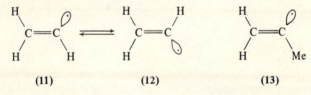

(11)   (12)   (13)

Chemical studies on transfer reactions of substituted vinyl radicals have also been used to give evidence about their structure. Decompositions of *cis*-

and *trans*-dicinnamoyl peroxides in bromotrichloromethane give both *cis*- and *trans*-β-bromostyrenes in proportions which reflect the stereochemistry of the initial peroxide (reactions *1* and *2*). This can be explained by postulating that scavenging of the bent radical by bromotrichloromethane competes with inversion of the radical (reactions *3* and *4*).

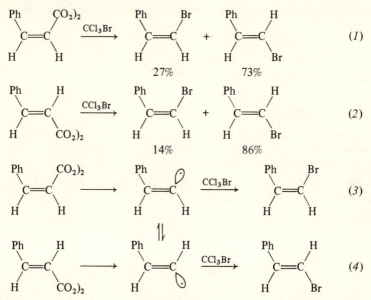

The only examples of vinyl radicals with linear structures are (14) (R = Bu^t or SiMe$_3$) i.e. very sterically congested radicals.

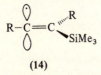

(14)

5.7. Group IV radicals

Unlike alkyl radicals, silyl radicals R$_3$Si $\cdot$ and germyl R$_3$Ge $\cdot$ radicals are non-planar. Chemical studies on silyl radicals generated from asymmetric silicon compounds indicate that they react with retention of configuration, i.e. silyl radicals have a reasonably high degree of configurational stability.

5.8. Nitrogen-centred radicals

(*a*) *Dialkylaminyl radicals* (R$_2$N $\cdot$). ESR studies indicate that the nitrogen in dialkylaminyl radicals is sp^2-hybridized and the unpaired electron is in a *p*-orbital: dialkylaminyl radicals are thus bent rather than linear.

(*b*) *Nitroxides* ($R_2\ddot{N}-O\cdot \leftrightarrow R_2\overset{+}{N}-O^-$). X-ray crystallographic studies indicate that many though not all dialkyl nitroxides have a non-planar arrangement of atoms about the nitrogen of the nitroxide group.

(*c*) *Iminoxy or oximino radicals*. The relatively high value of the nitrogen splitting constant for iminoxy radicals indicates that the unpaired electron is in an *s*-containing orbital. These radicals can therefore best be represented by

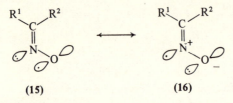

(15)   (16)

the canonical structures (15) and (16) in which the orbital containing the unpaired electron is perpendicular to the plane of the C=N—O system. Inversion about the nitrogen in iminoxy radicals is relatively slow on the ESR time scale though it is much too fast to enable the isomeric radicals (17) and (18) to be obtained free from each other.

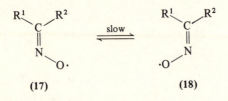

(17)   (18)

5.9.   Phosphorus-centred radicals

(*a*) *Phosphino radicals* ($R_2 P \cdot$). Phosphino radicals like amino radicals are bent with the unpaired electron in a *p*-orbital.

(*b*) *Phosphinyl radicals*. Both ESR and chemical studies (see reaction *5*) indicate that phosphinyl radicals, e.g. (19), are pyramidal and that they have a fair degree of configurational stability.

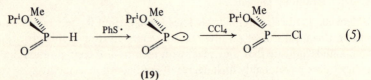

(19)

(*c*) *Phosphoranyl radicals*. Phosphoranyl radicals, $R_4 P \cdot$, which are important intermediates in radical displacement reactions on phosphorus (see p. 187), normally adopt the trigonal bipyramidal structure (20) in which the unpaired electron is in an orbital occupying an equatorial position rather than the alternative trigonal bipyramidal structure (21) with the unpaired electron in an apical (axial) position or the tetrahedral structure (**22**). The most electrophilic ligands tend to occupy the apical positions. The radicals have a fair

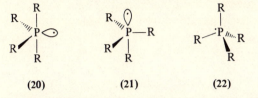

(20) (21) (22)

degree of configurational stability: there is little tendency for substituents in the apical and equatorial positions to interchange. Arylphosphoranyl radicals, however, tend to prefer the tetrahedral structure.

6 Radicals of long life

6.1. Introduction

In this chapter we shall be concerned with those radicals which have lifetimes of several minutes or greater in dilute solution in inert solvents. They can thus be studied by conventional spectroscopic techniques. Such radicals are referred to as long-lived or persistent radicals in contrast to transient radicals, such as the methyl radical, which are very much more reactive.

It is pertinent to draw a distinction between the lifetime or persistence of a radical and its stability. Stability is a thermodynamic property and is measured in terms of the difference in the $C-H$ bond strength of the radical in question and the $C-H$ bond strength of the appropriate alkane (primary, secondary or tertiary), while lifetime or persistence is a kinetic or reactivity property. The stabilization energy of the benzyl radical can thus be defined as $D(PhCH_2-H) - D(CH_3CH_2-H) = 54$ kJ mol^{-1}. Delocalization of unpaired spin which results in increased radical stability as measured by the decrease in the $C-H$ bond strength frequently has relatively little influence on the lifetime of radicals (see p. 8). This is certainly so for the benzyl radical (1) which in terms of reactivity is a highly reactive radical even though there is considerable delocalization of unpaired spin into the benzene ring (the spin density on the aliphatic carbon is about 0.7). The lifetime of a radical is much more profoundly influenced by steric shielding of the radical centre by bulky substituents which inhibit both radical–radical reactions and reactions of the radical with the solvent or other substrates. These points will be illustrated in this chapter from a consideration of several classes of what were once referred to as stable radicals but are now more correctly designated as long-lived or persistent radicals.

6.2. Triarylmethyl radicals

The best-known stable radical is triphenylmethyl which was discovered by Gomberg in 1900 whilst he was attempting to prepare hexaphenylethane. He obtained the colourless hydrocarbon (2), which he believed to be hexaphenyl-ethane (3), and which gave yellow solutions in a variety of solvents. The intensity of the colour increased on heating and was discharged by bubbling

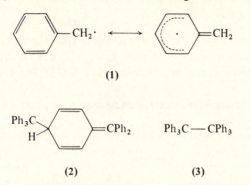

(1)

Ph₃C⟨⟩=CPh₂   Ph₃C—CPh₃

(2)   **(3)**

oxygen into the solution. These solutions failed to obey Beer's law. This evidence indicated that free triphenylmethyl radicals were in equilibrium with their dimers. This was confirmed initially by molecular weight determinations and later by ESR spectroscopy.

The ESR spectra of triphenylmethyl radicals indicate that there is con-siderable delocalization of the unpaired electron onto the *ortho-* and *para*-positions of the phenyl rings (see (4)). The extent of delocalization of the unpaired electron is not, however, very much greater for triphenylmethyl radicals than for diphenylmethyl or benzyl radicals. The greater lifetime of

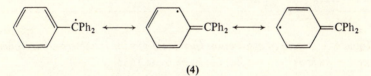

(4)

triphenylmethyl radicals must therefore be due more to steric influences which prevent dimerization rather than delocalization of the unpaired elec-tron. That the unpaired electron is not much more extensively delocalized in the triphenylmethyl radical than in the benzyl radical arises from the lack of coplanarity of the phenyl groups in the former. X-ray crystallographic studies have shown that the benzene rings are tilted at an angle of 30° in the stable tris(*p*-nitrophenyl)methyl radical, whilst ESR studies indicate that the aryl rings are tilted at an angle of 50° in the tris(2,6-dimethoxyphenyl)methyl radical.

The dimer (2) arises from attack by the trigonal carbon atom of one tri-phenylmethyl radical at the *para*-position of one of the phenyl groups in a second radical. Simple C—C coupling of triphenylmethyl radicals to give hexa-phenylethane does not occur as there would be very severe steric interaction between the phenyl groups in the approaching radicals.

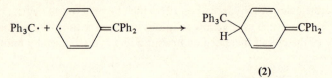

$$(2)$$

The importance of steric effects upon the stability of triarylmethyl radicals with respect to their dimers is clear from a study of the effect of substituents on the degree of dissociation of the dimer (tables 6.1, 6.2, 6.3). These results

Table 6.1. *Degree of association (per cent) of triarylmethyl radicals.* (0.2 M solutions in benzene.)

Ar	$ArPh_2C\cdot$	$Ar_2PhC\cdot$	$Ar_3C\cdot$
$o\text{-MeC}_6H_4$	75	18	13
$m\text{-MeC}_6H_4$	93.5	93	60
$p\text{-MeC}_6H_4$	95	94.5	84

Table 6.2. *Degree of association (per cent) of some tris(o-halophenyl)methyl radicals.* (0.2 M solutions in benzene.)

	X = F	Cl	Br
Association of $(o\text{-}XC_6H_4)_3C\cdot$	92.5	88	83

Table 6.3. *Degree of association (per cent) of some arylbis(p-t-butylphenyl)-methyl radicals.* (0.1 M solutions in benzene.)

	Ar = Ph	$o\text{-MeC}_6H_4$	$o\text{-BrC}_6H_4$	$p\text{-Bu}^tC_6H_4$
Association of $Ar\dot{C}(C_6H_4-Bu^t-p)_2$	92.5	35	6	0

enable the following conclusions to be drawn:
(a) The degree of dissociation of the radical is increased more by substitution at the *ortho-* than at the *meta-* and *para-*positions.
(b) The degree of dissociation of the radical increases with the number and size of the *ortho-*substituents.
(c) The degree of dissociation of the radical is very high if all three rings are substituted in the *para-*position by bulky substituents.

An increasing amount of *ortho-*substitution increases the frontal strain in the radical, thereby inhibiting attack on a second radical. Substitution in all three *para*-positions of the triphenylmethyl radical by bulky groups renders attack by a second radical impossible. If, however, there is one free *para*-position the degree of dissociation is markedly reduced. These results were not explicable with the hexa-arylethane formulation of the dimer.

In recent years a number of perchloro radicals such as perchlorotriphenyl-methyl and perchlorodiphenylmethyl have been synthesized. These have much longer lifetimes than even the most persistent of the radicals mentioned previously: they do not react with either oxygen or molecular chlorine. Per-chlorotriphenylmethyl can be recovered unchanged from boiling toluene and does not decompose appreciably at 300°C. A model of this radical indicates that the aromatic rings are tilted at 60° to each other and thus there is little or no delocalization of the unpaired electron. This is confirmed by its ESR spectrum consisting of a single line which indicates that the unpaired electron is not delocalized. The model also indicates that the methyl carbon is almost completely shielded by the perchlorophenyl rings.

The above results indicate that steric effects are much more important than delocalization effects in determining the persistence of triarylmethyl radicals. This is supported by consideration of the 9-phenylfluorenyl radical (5) which is almost completely associated. In this radical the degree of delocal-ization of the unpaired electron is considerably enhanced on account of its planarity, but at the same time there is a marked decrease in steric effects which allows dimerization to occur. In contrast, the related 9-mesitylfluorenyl

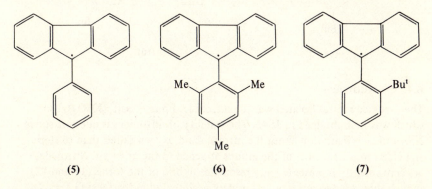

(5)　　　　　　　　(6)　　　　　　　　(7)

(6) and 9-(*o*-t-butylphenyl)fluorenyl radicals (7) are completely dissociated as steric effects prevent dimerization.

6.3.　Phenoxy radicals

The phenoxy radical (8) is a much more persistent radical than an alkoxy radical, on account of the delocalization of the unpaired electron onto the *ortho*- and *para*-positions of the benzene ring. Coupling of the delocalized phenoxy radical produces dimers which are formed by C—O or C—C but not

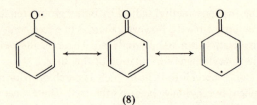

(8)

O—O combination (see p. 141). O—O Combination does not take place, on account of the instability of the resultant peroxide. In the case of C—O and C—C dimerization this may take place at either the *ortho-* or *para-*positions. Thus, when the *ortho-* and *para-*positions are substituted by bulky groups there will be steric hindrance to dimerization, and consequently 2,4,6-trisubstituted phenoxy radicals are much more persistent than the phenoxy radical itself. The 2,4,6-tri-t-butylphenoxy radical (9) is completely mono-meric both in solution and in the solid state. Galvinoxyl (10), in which the degree of electron delocalization is greater, is even more stable than the 2,4,6-tri-t-butylphenoxy radical. Galvinoxyl is extensively used as a radical scavenger.

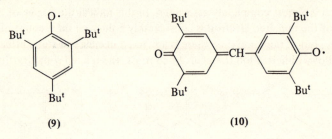

(9)   (10)

6.4.   Nitroxides

The first free radical isolated was the nitroxide, Fremy's salt, $(KSO_3)_2 NO \cdot$, which was first obtained in 1845. The stability of nitroxides is due to a stable electronic configuration about the nitrogen and oxygen rather than to steric and electronic influences of the groups attached to the nitrogen. Nitroxides are conveniently represented as a resonance hybrid of the forms (11 and 12), though they are perhaps more accurately represented in the form (13) in which there is a three-electron bond between the nitrogen and the oxygen. As a consequence of this stable electronic arrangement they do not undergo dimerization at the nitrogen or oxygen atoms.

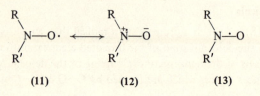

(11)   (12)   (13)

The lifetime of nitroxides is largely determined by their ability to react with themselves by disproportionation. Nitroxides persist when the groups attached to nitrogen do not allow the radical to undergo reaction with itself. Thus, whilst dialkyl nitroxides (14) in which either alkyl group is primary or secondary are extremely short-lived because of the ease with which they undergo disproportionation to hydroxylamine (15) and nitrone (16), di-t-alkyl nitroxides are quite persistent. (This is due more to their inability to undergo reaction because of the lack of a mechanism whereby they may react, rather than to steric shielding of the $>$N—O$\cdot$ group.)

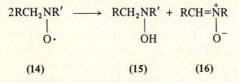

(14)   (15)   (16)

The groups attached to the nitrogen may make the radical more susceptible to dimerization at some other part of the radical by increasing the amount of electron delocalization, and hence the number of reactive sites. Thus, t-butyl phenyl nitroxide is much less stable than di-t-butyl nitroxide, because the unpaired electron is delocalized onto the aromatic nucleus. As a result of this extra delocalization of the unpaired electron, the nitrogen coupling constant is lower for t-butyl phenyl nitroxide (1.34 mT) than for di-t-butyl nitroxide (1.62 mT). Chemically this delocalization makes the aromatic nucleus susceptible to attack by a second radical at the *para*-position (scheme 1). The

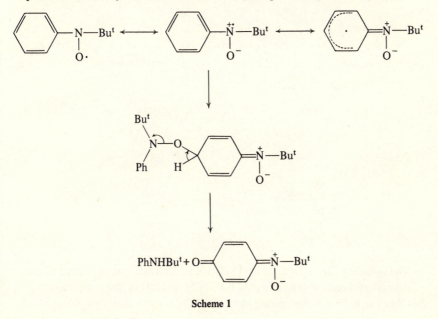

Scheme 1

resulting dimer immediately breaks down to *N*-t-butylaniline and *N*-t-butyl-*p*-benzoquinonimine-*N*-oxide. Carbon–carbon dimerization does not occur as is the case for phenoxy radicals, because most of the electron density is associated with the nitrogen and oxygen atoms.

6.5.   Hydrazyl radicals

Possibly the best known long-lived organic radical is the 2,2-diphenyl-1-picrylhydrazyl radical. It is quite stable in the solid state and also in dilute solution. Its importance lies in its extensive use in polymer chemistry as a radical scavenger, and to a lesser extent as a standard in ESR spectroscopy for measuring radical concentrations. The ESR spectrum of the 2,2-diphenyl-1-picrylhydrazyl radical indicates that the unpaired electron is associated almost equally with both nitrogens; the nitrogen coupling constants with N-1 and N-2 being 0.935 and 0.785 mT respectively. There is also appreciable delocalization of the unpaired electron onto the aromatic residues. Six canonical forms can be considered (**17–22**).

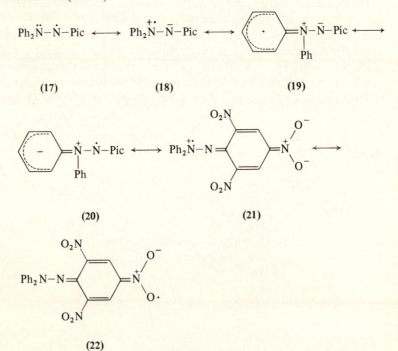

The tendency for triarylhydrazyl radicals to dimerize can be related to the relative importance of the canonical forms (**23 and 24**) to the resonance hybrid. The greater is the contribution of (**23**) the more dimerization will be

$$Ar_2\ddot{N}-\dot{N}Ar' \longleftrightarrow Ar_2\overset{+\cdot}{N}-\overset{\cdot\cdot}{N}Ar'$$

(23)   (24)

favoured. Electron-withdrawing substituents in the Ar group and electron-releasing substituents in the Ar′ group will favour the dipolar canonical form (24) and hence increase the lifetime of the radical. Other radicals, e.g. (25)–(27) are similarly stabilized by simultaneous conjugation with electron-donor and electron-acceptor groups.

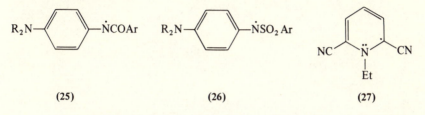

(25)   (26)   (27)

6.6.   Sterically hindered alkyl, aryl, vinyl and allyl radicals

Steric shielding of the radical centre as occurs in the perchlorotriphenylmethyl radical results in exceptionally low reactivity. Recently a number of long-lived alkyl radicals, e.g. $Bu_3^t C\cdot$, $(Me_3Si)_3C\cdot$, and $Bu_2^t CH\cdot$ have been prepared: these radicals cannot dimerize readily because of steric hindrance, nor can they disproportionate. The di-t-butylmethyl radical decays by a first-order β-scission.

$$Bu_2^t \dot{C}H \rightarrow Bu^t CH=CMe_2 + Me\cdot$$

The long lifetimes of a series of 1,1,2,2-tetrasubstituted ethyl radicals, $(Me_3Si)_2\dot{C}CH(SiMe_3)_2$, $Bu^t(Me_3Si)\dot{C}CH(SiMe_3)_2$, $Bu_2^t \dot{C}CH(SiMe_3)_2$ and $Bu_2^t \dot{C}CHBu_2^t$, all of which might be expected to undergo disproportionation, are attributed to the fact that they are locked in the conformation (28) in

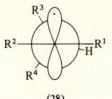

(28)

which the lone carbon–hydrogen bond is at right angles to the *p*-orbital containing the unpaired electron: disproportionation would occur much more readily if the C—H bond undergoing hydrogen transfer was in the plane of the orbital containing the unpaired electron as the developing *p*-orbital on the β-carbon would overlap the *p*-orbital containing the unpaired electron.

Steric shielding of the radical centre has also been used to enhance the lifetimes of aryl and vinyl radicals. Thus the radicals (29), (30) (R = Bu^t or Me$_3$Si) and (31) (R = Bu^t or Me$_3$Si) all have appreciable lifetimes. The lifetime of allyl radicals is similarly increased by bulky groups as in (32) even though this substitution may result in the radicals being twisted out of planarity with consequent reduction in delocalization of the unpaired spin.

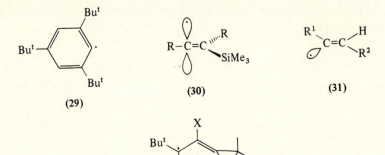

(29)

(30)

(31)

(32)

7 Radical–radical reactions

7.1. Introduction

There are two basic ways in which radicals can react with each other, namely *combination* and *disproportionation*. When two *like* radicals, such as phenyl, couple together the process is called dimerization:

$$Ph \cdot + Ph \cdot \rightarrow Ph-Ph$$

This is one of the most characteristic radical reactions which virtually all carbon-centred radicals and heteroradicals undergo. If dimers are detected in a reaction mixture this is an indication that the mechanism may involve radical intermediates. The alternative mode of reaction of radicals with each other involves transfer of an atom, usually hydrogen or halogen, from one radical to the other giving a saturated and an unsaturated product, e.g.:

$$CH_3 CH_2 \cdot + CH_3 CH_2 \cdot \rightarrow CH_3 CH_3 + CH_2 = CH_2$$

In cross-disproportionation involving two *unlike* radicals a hydrogen can be transferred to either of the radicals so that two sets of products are formed, e.g.:

$$CH_3 CH_2 \cdot + CH_3 CH_2 CH_2 \cdot \rightarrow CH_3 CH_3 + CH_3 CH = CH_2$$

$$CH_3 CH_2 CH_2 \cdot + CH_3 CH_2 \cdot \rightarrow CH_3 CH_2 CH_3 + CH_2 = CH_2$$

Radical-chain reactions are terminated by combination or disproportionation of the radicals involved. If the chains are long the radical–radical products form a very small part of the total and may be undetectable. Dimers and disproportionation products can be the main reaction products if the radical concentration is very high, as in anodic oxidation reactions, or where there are no reactive substrates or solvents present. Radical combination is important in kinetic and mechanistic studies of radical reactions as a reference reaction. The rates of other reactions occurring in the system can be measured relative to that of the dimerization of the primary radicals. Considerable effort has gone into the measurement of absolute rate constants for radical combination reactions so that accurate absolute rate constants can be deduced for other processes.

7.2.   Atom combination reactions

When two atoms combine, the energy produced in forming the bond is also sufficient to break the bond again. The atoms lack vibrational and rotational degrees of freedom to absorb this energy so they tend to fall apart again before they can be stabilized by collision with other molecules of gas. There are three main conditions under which this energy can be transferred from the diatomic product so that combination is completed. For a number of atomic species such as halogen atoms the excess energy can be removed from the combination product by emission of light or infrared radiation.

$$A\cdot + A\cdot \rightarrow A_2 + h\nu$$

Atom combination can also occur as a third-order process through the intermediacy of a 'third body' or 'chaperon' molecule (M) which may be added to the system and absorbs the excess energy.

$$A\cdot + A\cdot + M \rightarrow A_2 + M$$

The energy produced in the reaction appears as translational, vibrational and rotational energy of the chaperon. Thirdly, the wall of the reaction vessel can remove the excess energy and so catalyse the combination reaction. The third-order and wall combinations are also frequently accompanied by emission of radiation.

The rates of the third-order homogeneous combination reactions depend on the nature and pressure of the third body present. The product, diatomic molecule A_2, is itself usually a very efficient catalyst for the reaction. This is particularly true for iodine and bromine atoms where the molecules I_2 and Br_2 are about 500 and 20 times more effective respectively than argon as chaperons.

For the majority of atoms the wall combination process is first order in the atom concentration, but some atoms, such as hydrogen and nitrogen also seem to take part in second-order wall processes:

$$A\cdot + Wall \rightarrow \tfrac{1}{2}A_2 + Wall$$

$$A\cdot + A\cdot + Wall \rightarrow A_2 + Wall$$

Pyrex and quartz surfaces are relatively inefficient at catalysing this reaction and wall combination is not important in reaction vessels of these materials except at low pressures (less than 5 torr) in the gas phase. Metals such as copper, silver or platinum on the other hand, are very highly active for atom combinations.

In solution, atom combination is second order because of the great excess of solvent available to remove the energy of the reaction. The reaction is very fast and takes place at the maximum possible rate, which is the rate at which the atoms can diffuse together in the medium. Consequently, the combination of atoms such as bromine or iodine is said to be 'diffusion controlled' in solution.

7.3.   Combination and disproportionation of alkyl radicals

When two polyatomic radicals combine, the energy released in forming the bond can be shared out amongst the vibrational and rotational modes of the dimeric product so that it does not immediately fall apart again. The dimer has a sufficiently long lifetime to allow it to be stabilized by collision with other molecules, except at low pressures in the gas phase where collisions are rare events. The combination reactions of almost all radicals containing more than two atoms are therefore second-order processes in the gas phase and in solution.

Combination rate constants have been measured in the gas phase and in solution by the intermittent illumination (rotating sector) method, or by methods which directly measure the radical concentration such as kinetic spectroscopy, electron spin resonance spectroscopy or mass spectrometry. For carbon-centred radicals in the gas phase the rate constants are usually in the range 10^8 to 10^{11} $1\,mol^{-1}\,s^{-1}$ and are virtually independent of temperature and pressure. The following order of reactivity has been found:

$$Me \cdot > Et \cdot > Pr^i \cdot > Bu^t \cdot$$

The lower rate of reaction of the larger, more crowded radicals can probably be attributed to steric inhibition of the approach of one radical to the other. Di- and tri-t-butylmethyl radicals, and other radicals with bulky substituents at the radical centre, are exceptionally long-lived on account of their inability to disproportionate and because of steric hindrance to dimerization (see p. 53).

In a chemical system containing two unlike radicals there are three possible combination products. For example, methyl radicals may be produced by decomposition of di-t-butyl peroxide in carbon tetrachloride solution. The methyl radicals will then abstract chlorine from the solvent giving trichloromethyl radicals.

$$(CH_3)_3CO \cdot OC(CH_3)_3 \rightarrow 2CH_3 \cdot + 2(CH_3)_2CO$$

$$CH_3 \cdot + CCl_4 \rightarrow CCl_3 \cdot + CH_3Cl$$

Methyl and trichloromethyl radicals in this mixture can then dimerize to give ethane and hexachloroethane respectively, or cross-combination can occur giving trichloroethane:

$$CH_3\cdot\ +\ CH_3\cdot\ \xrightarrow{\ k_{11}\ }CH_3CH_3$$

$$CH_3\cdot\ +\ CCl_3\cdot\ \xrightarrow{\ k_{12}\ }CH_3CCl_3$$

$$CCl_3\cdot\ +\ CCl_3\cdot\ \xrightarrow{\ k_{22}\ }CCl_3CCl_3$$

The kinetic analysis of cross-combination reactions is usually made in terms of the cross-combination rate constant ratio; $k_{12}/(k_{11}k_{22})^{1/2}$. Simple collision theory predicts that this ratio should be about 2 for reactions in the gas phase, and this has been confirmed by experiment for a large number of alkyl and related radicals. This ratio is sometimes found to be close to 2 in solution as well, but wide divergences from 2 have also been observed for some radicals such as the bulky species involved in polymerization.

Disproportionation is also a very fast second-order process in the gas phase and in solution. This reaction is not a straightforward hydrogen abstraction from one radical by the other. Hydrogen abstraction reactions have rates slower by a factor of about ten than disproportionations, and hydrogen abstractions have activation energies of, typically, 40 kJ mol^{-1} whereas the disproportionation step is temperature independent in the gas phase. The ethylene formed in the disproportionation of two $\alpha\alpha$-dideuterioethyl radicals, was shown to be exclusively 1,1-dideuterioethylene:

$$CH_3CD_2\cdot + CH_3CD_2\cdot\ \longrightarrow\ CH_3{-}CD_2\overset{H}{\underset{\cdot CD_2}{\diagup}} CH_2\ \longrightarrow\ CH_3CD_2H + CD_2{=}CH_2$$

Thus the transferred hydrogen comes wholly or mainly from the β-carbon atom of the radical.

Disproportionation is slower than combination for primary and secondary carbon-centred radicals in the gas phase or in solution, and combination products predominate for most radicals. The disproportionation/combination rate constant ratio, k_d/k_c, lies between 0.5 and 0.05 for many radicals. However, disproportionation is favoured for radicals having many hydrogen atoms in the β-position, and k_d/k_c increases roughly linearly with the number of β-hydrogens until for t-butyl radicals containing nine β-hydrogens it is greater than one. Radicals such as $Bu_2^t\dot{C}CH_2Me$, with bulky substituents at the radical centre are very unreactive and do not disproportionate even though β-hydrogens are available. In the preferred conformation of this kind of radical the dihedral angle between the p-orbital containing the unpaired electron and the β-C$-$H bond is 60° or more, so that formation of a double bond by transfer of the β-hydrogen is energetically unfavourable (see p. 53). The stereoelectronic requirements for the disproportionation reactions are clearly quite strict.

7.4. Cage reactions and diffusion control in solution

In solution each radical is closely surrounded by solvent molecules, and two radicals must diffuse together before they can combine. Once they have encountered each other, however, they will be constrained by the 'cage' of solvent molecules to undergo several collisions before they diffuse apart again. For a fast reaction, such as combination which occurs very rapidly once the radicals have encountered each other, the probability is very high that each encounter in solution will lead to reaction. The reaction will therefore take place at the rate at which the radicals can diffuse together. This kind of reaction is said to be 'diffusion controlled', and the diffusion rate is the maximum rate of combination in solution.

The rate of diffusion is given approximately by:

$$k_{(diff)} = 8RT/3000\eta$$

and depends inversely on the viscosity coefficient, η, of the solvent. Diffusion rates are generally of the order 10^8 to 10^9 $1 \, mol^{-1} \, s^{-1}$ at room temperature. There is good experimental evidence that alkyl radical combination reactions are diffusion controlled in solution. The measured rate constants for combination are usually around 10^9 $1 \, mol^{-1} \, s^{-1}$, close to the diffusion limit, and are found to depend very little on the structure of the radical. Similarly, the combination rate constants of alkyl radicals vary inversely with the viscosity of the medium.

When two radicals are generated together in close proximity, as happens when an azo-compound or a peroxide is decomposed, the cage effect of the solvent will cause them to undergo several collisions before they can diffuse into the bulk of the solution. There will thus be a strong probability that the two radicals never separate, but combine before they can diffuse apart. This cage effect thus leads to a preponderance of the products formed from radical-radical reactions of the initial pair. For example, in the decomposition of an unsymmetrical azo-compound in solution:

$$EtN_2Me \xrightarrow{\Delta} [Et \cdot N_2 \cdot Me]_{cage} \xrightarrow{cage} n\text{-}C_3H_8$$

$$\downarrow diffusion$$

$$C_2H_6 + n\text{-}C_4H_{10}$$

the methyl and ethyl radicals will combine in the solvent cage to give propane and this will predominate over the combination products, ethane and butane, which can only form from those radicals which escape from the solvent cage. Similarly, if two different symetrical azo-compounds are decomposed in the same solution the cage combination products will be the symetrical dimers, and these will predominate over the cross-combination product which can only form when the radicals diffuse out of the cage into the bulk of the solution.

The amount of cage combination depends on the solvent, and cage reaction forms a greater percentage of the total combination in more viscous solvents. In very viscous solvents such as decalin, cage combination may account for as much as 80 per cent of the total reaction, as compared with only 15 per cent in fluid solvents like pentane. The amount of cage combination also decreases with increasing temperature.

The possibility that radicals generated close together within a solvent cage might combine stereospecifically, with retention of their initial configuration, was investigated by decomposing an optically active azo-compound such as (S)-(−)-1,1′-diphenyl-1-methylazomethane (**1**).

$$\underset{\text{Me}}{\overset{\text{Ph}}{\diagdown}}\text{C}-\text{N}=\text{N}-\text{CH}_2\text{Ph} \quad \xrightarrow[\text{C}_6\text{H}_6]{\Delta} \quad \left[\underset{\text{Me}}{\overset{\text{Ph}}{\diagdown}}\text{C}\cdot \quad \text{N}_2 \quad \cdot\text{CH}_2\text{Ph}\right]_{\text{cage}} \quad \longrightarrow \quad \underset{\text{Me}}{\overset{\text{Ph}}{\diagdown}}\text{C}-\text{CH}_2\text{Ph}$$

(1)   (2)

If inversion of configuration were slow in comparison with cage combination, then the product (**2**) would be formed with retention of configuration and consequent optical activity. In fact only about 15 per cent retention of configuration was observed, showing that *rotation* of the radicals in the solvent cage is much faster (about 16 times faster) than combination.

7.5. Combination and disproportionation of heteroradicals

Heteroradicals, in which the unpaired electron is localized mainly on an atom other than carbon, also react with each other by the processes of combination and disproportionation which are so characteristic of any radical system. The radical–radical behaviour of heteroradicals is governed mainly by the tendency of the heteroatoms to form double bonds, which influences the extent of disproportionation, and by the strength of the bond formed between the two heteroatoms in the combination product.

The main radical–radical reaction for silyl, germyl and stannyl radicals is combination because of the unstable nature of the double bond to these heteroatoms. Alkylsilyl and alkyltin radicals combine rapidly and efficiently to give the corresponding disilane and ditin derivatives, e.g.:

$$\text{Ph}_3\text{Si}\cdot + \text{Ph}_3\text{Si}\cdot \rightarrow \text{Ph}_3\text{SiSiPh}_3$$

$$\text{Bu}^n_3\text{Sn}\cdot + \text{Bu}^n_3\text{Sn}\cdot \rightarrow \text{Bu}^n_3\text{SnSnBu}^n_3$$

Disproportionation does occur between an alkyl radical and a silyl or tin radical e.g.:

$$Me_3Si \cdot + Me_3C \cdot \rightarrow Me_3SiH + Me_2C{=}CH_2$$

Nitrogen radicals both dimerize and disproportionate. In particular, alkyl-aminyl radicals such as diethylaminyl give substituted hydrazines by combination and imines by disproportionation:

$$2Et_2N \cdot \rightarrow Et_2NNEt_2$$

$$2Et_2N \cdot \rightarrow EtN{=}CHCH_3 + Et_2NH$$

Phosphinyl radicals and alkylphosphinyl radicals disproportionate rather than dimerize, to give phosphenes,

$$\cdot PH_2 + \cdot PH_2 \rightarrow PH_3 + : PH$$

$$: PH + : PH \rightarrow P_2 + H_2$$

and these then react to give phosphorus and hydrogen. Diaryl- and dihalo-phosphinyl radicals, however, react by combination.

Alkoxy radicals react by both combination and disproportionation. The oxygen–oxygen bond in the combination product is very weak, whereas the oxygen–hydrogen bond formed on disproportionation is very strong. Consequently the disproportionation mode is very much favoured for simple alkoxy radicals; k_d/k_c being of the order of ten. Alkylthiyl radicals combine to give disulphides, and disproportionate to give a mercaptan and a thioketone (or aldehyde) e.g.:

$$2Me_2CHS \cdot \rightarrow Me_2CHSSCHMe_2$$

$$2Me_2CHS \cdot \rightarrow Me_2CHSH + Me_2C{=}S$$

8   Radical transfer reactions

8.1.   The kinetic problem

The most familiar reactions of radicals are those in which they abstract an atom, usually a hydrogen atom from a neutral molecule, to give a new radical and a new stable molecule:

$$CF_3 \cdot + CH_4 \rightarrow CF_3 H + CH_3 \cdot$$

Probably the most common example of a hydrogen abstraction is halogenation, especially chlorination:

$$Cl \cdot + C_2 H_6 \rightarrow HCl + C_2 H_5 \cdot$$

Although the majority of radical transfers involve hydrogen abstraction, other atoms can be abstracted, especially halogen atoms:

$$CCl_3 Br + C_3 H_7 \cdot \rightarrow C_3 H_7 Br + CCl_3 \cdot$$

We have seen (chapter 1) that in reactions involving reactive radicals they normally have only a transient existence and their concentration is usually very low. High conversion in a chlorination reaction is achieved because a chain reaction is involved. We can depict a chlorination reaction as follows:

$$Cl_2 \xrightarrow{h\nu} 2Cl \cdot \qquad\qquad \textit{initiation}$$

$$\left. \begin{aligned} Cl \cdot + CH_4 \xrightarrow{(2)} HCl + CH_3 \cdot \\[2mm] CH_3 \cdot + Cl_2 \xrightarrow{(3)} CH_3 Cl + Cl \cdot \end{aligned} \right\} \quad \textit{chain propagation}$$

$$\left. \begin{aligned} Cl \cdot + Cl \cdot + M \xrightarrow{(4)} Cl_2 + M \\[2mm] CH_3 \cdot + Cl \cdot \xrightarrow{(5)} CH_3 Cl \\[2mm] CH_3 \cdot + CH_3 \cdot \xrightarrow{(6)} C_2 H_6 \end{aligned} \right\} \quad \textit{termination processes}$$

The termination processes, depending as they do on the product of two very low concentrations, are comparatively slow so that the chain propagating steps are repeated many times before the chain terminates in one of the reactions (4), (5) and (6). The rate-determining propagation step is extremely difficult to measure.

$$\text{Rate (2)} = V_2 = k_2 [\text{Cl} \cdot] [\text{CH}_4]$$

There is no easy means of determining the concentration of chlorine atoms, even if we know the rate of formation of chlorine atoms, since they are also formed in reaction (3) and lost in reactions (4) and (5). Chain reactions of this kind all present this problem and measurement of the relevant rate constants usually depends on using the 'steady-state approximation'. If we have a bulb containing chlorine and methane in the dark, when we shine a light into the vessel chlorine atoms will be formed and the reaction will start. At the very beginning, each molecule of chlorine dissociated will lead to two new chains and the number of reaction chains will increase until the concentration of atoms and radicals is sufficient for the number of termination reactions to become appreciable. Eventually, unless the light source is of enormous power, a stage will be reached when the rate of termination will equal the rate of initiation and a steady state will have been reached. In the steady state radicals are being formed as fast as they are being destroyed by reaction. We can express this mathematically:

$$\frac{d[\text{CH}_3 \cdot]}{dt} = 0 = k_2 [\text{Cl} \cdot] [\text{CH}_4] - k_3 [\text{CH}_3 \cdot] [\text{Cl}_2] - k_5 [\text{CH}_3 \cdot] [\text{Cl} \cdot]$$
$$- k_6 [\text{CH}_3 \cdot]^2$$

We can write a similar expression for chlorine atoms, provided we have an expression for the rate of formation by photolysis.

$$\text{Rate (1)} = \phi I_a$$

ϕ is the quantum yield of chlorine molecule decomposition and I_a is the light absorbed. With this expression for the rate of the initiation step we can now express the steady state condition again:

$$\frac{d[\text{Cl} \cdot]}{dt} = 2\phi I_a - k_2 [\text{Cl} \cdot] [\text{CH}_4] + k_3 [\text{CH}_3 \cdot] [\text{Cl}_2] - k_4 2 [\text{Cl} \cdot]^2 [\text{M}]$$
$$- k_5 [\text{CH}_3 \cdot] [\text{Cl} \cdot] = 0$$

Usually we can neglect reactions (5) and (6) compared with reaction (4) where M, the third body, includes the wall of the reaction vessel. With this simplification we have:

$$[\text{Cl} \cdot] = \left(\frac{\phi I_a}{k_4 [\text{M}]} \right)^{1/2}$$

Thus the rate of chlorination (reaction (2)) is given by

$$\text{Rate (2)} = k_2 \left(\frac{\phi I_a}{k_4 [M]} \right)^{\frac{1}{2}} [CH_4]$$

We still have problems, for although ϕI_a can be determined experimentally quite easily, k_4 is a much more difficult problem (determination of the absolute rate of atom and radical combination reactions is discussed in chapter 7). However, if the rate of one chlorination reaction can be determined then the rates of other chlorination reactions can be obtained by competitive methods. If two hydrocarbons RH and R'H are competitively chlorinated the ratio of the rate of formation of RCl to the rate of formation of R'Cl is given by the following equation:

$$\frac{\text{Rate RCl}}{\text{Rate R'Cl}} = \frac{k_2}{k_2'} \times \frac{[RH]}{[R'H]}$$

If the total conversion is small so that the ratio of [RH] to [R'H] has not changed, integration from the initial state when there is no RCl or R'Cl to the final state gives

$$\frac{[RCl]_f}{[R'Cl]_f} \approx \frac{k_2 [RH]_i}{k_2' [R'H]_i}$$

where subscripts f and i, denote the initial and final states respectively, hence

$$\frac{k_2}{k_2'} = \frac{[RCl]_f [R'H]_i}{[R'Cl]_f [RH]_i}.$$

When different sites in the same molecule are being studied, the ratio $[R'H]_i : [RH]_i$ is simply the ratio of the number of hydrogen atoms of each type in the molecule. For instance if butane is chlorinated the rate of attack per hydrogen atom at the primary and secondary sites, known as the *Relative Selectivity (RS)*, is given by

$$RS_p^s = k_2^s / k_2^p \times 3/2$$

where subscripts and superscripts s and p, stand for secondary and primary respectively.

8.2. The thermodynamic problem

The simplest radical transfer reaction is that of a hydrogen atom with molecular hydrogen.

$$H \cdot + H_2 \rightarrow H_2 + H \cdot$$

This reaction, which has been the subject of much theoretical and experimental study, can be accelerated (like most other reactions) by raising the temperature. The rate of any such bimolecular reaction

$$X \cdot + HR \rightarrow XH + R \cdot$$

is given by

$$\text{Rate} = k[X \cdot][HR]$$

and the effect of temperature finds experimental expression in the Arrhenius equation

$$k = A \exp(-E/RT)$$

Where A is a constant with the same units as k, i.e. $dm^3 \, mol^{-1} \, s^{-1}$
E is the *Activation energy* in $kJ \, mol^{-1}$
R is the gas constant = 8.314 $J \, mol^{-1} \, K^{-1}$
T is the temperature in K

Let us consider a hydrogen abstraction involving the attack of a radical (or atom) $X \cdot$ on the molecule R—H.

$$X \cdot + H - R \longrightarrow X \cdots H \cdots R \longrightarrow X - H + R \cdot$$

(Reactants in the   (Activated complex   (Products in the
initial state)   in the transition state)   final state)

As $X \cdot$ approaches RH there will be increasing electronic repulsion until the electrons in the H—R bond become 'uncoupled' and the new bond between X and H begins to form. This will represent the top of a potential energy barrier and the intermediate state is the 'activated complex' usually denoted by the symbol ($\ddagger$). As the reaction continues, repulsion between the new molecule (HX) and R $\cdot$ will decrease as the products separate. We can visualize three types of reaction, (i) *exothermic*, where $D(H-X) > D(H-R)$ and the final state is of lower energy than the initial state, e.g. the fluorination of methane; (ii) *thermoneutral*, where $D(H-X) = D(H-R)$ and the initial and final states are of equal energy e.g. hydrogen atoms attacking molecular hydrogen and (iii) *endothermic*, where $D(H-X) < D(R-H)$, where the final state is of higher energy than the initial state, e.g. the bromination of methane. It is usual to represent these reactions by diagrams in which energy is plotted against the extent of reaction between a single atom $X \cdot$ and a single molecule RH. The important feature of these diagrams is that they show an energy barrier even when the reaction is extremely exothermic (e.g. the reaction of fluorine atoms with methane). The energy barrier, which can be related to the activation energy E of the Arrhenius equation, depends to some extent on the heat of reaction (ΔH), i.e. E for a very exothermic reaction will be small compared with that for an endothermic reaction; in the latter case E must be greater than ΔH and will be large. However, *there is in general no direct relationship between E and ΔH*. This can be illustrated by the reaction of hydrogen atoms with hydrogen chloride and with methane. Both reactions are very slightly exothermic and the first has an activation energy considerably *less* than that of the thermo-neutral reaction $H \cdot + H_2$, while the second has an activation energy which is significantly *greater*.

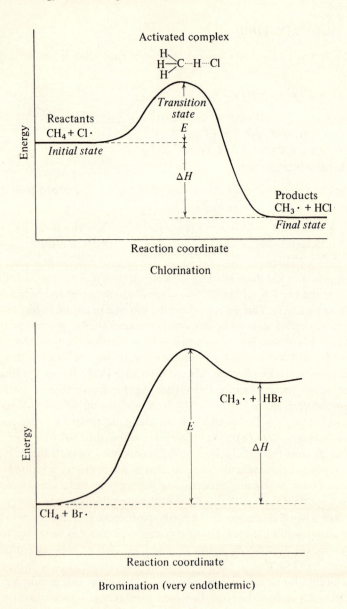

Chlorination

Bromination (very endothermic)

Figure 8.1 (caption on facing page).

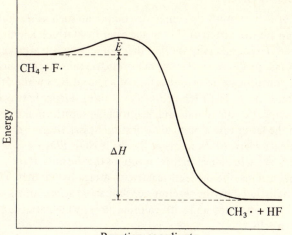

Reaction coordinate

Fluorination (very exothermic)

Figure 8.1. Diagrams showing energy states during the reaction of methane with atoms of chlorine, bromine and fluorine.

$$H\cdot + H_2 \rightarrow H_2 + H\cdot \qquad \begin{cases} \Delta H = 0 \\ E = 33 \text{ kJ mol}^{-1} \end{cases}$$

$$H\cdot + HCl \rightarrow H_2 + Cl\cdot \qquad \begin{cases} \Delta H = -4 \text{ kJ mol}^{-1} \\ E = 15 \text{ kJ mol}^{-1} \end{cases}$$

$$H\cdot + CH_4 \rightarrow H_2 + CH_3\cdot \qquad \begin{cases} \Delta H = -4 \text{ kJ mol}^{-1} \\ E = 48 \text{ kJ mol}^{-1} \end{cases}$$

Further discussion of the factors which control the activation energy of radical transfer reactions will be the subject of the next section.

The pre-exponential 'A' factor of the empirical Arrhenius equation, normally has a magnitude of about 10^{10} l mol^{-1} s^{-1} for reactions involving atoms and about 10^7 l mol^{-1} s^{-1} for reactions of simple radicals. This difference is predicted by 'transition-state' theory, which treats the 'activated complex' occurring at the top of the potential energy barrier like a molecule saving only in translation along the line of the reacting centres. According to this theory the 'A' factor of the Arrhenius equation is related to the entropy change in forming the activated complex. When a radical takes part in a transfer reaction three degrees of rotational freedom are lost, but this is not the case for an atom, so that the pre-exponential factor is smaller in the radical case. Although the pre-exponential term varies, the major changes in rate for the hydrogen abstraction by a particular atom or radical are usually associated with changes in the activation energy.

8.3.   Substituent effects

When an atom or radical attacks an aliphatic compound such as methyl pentanoate there are five different sites in the ester from which a hydrogen can be abstracted. Experimentally we find that not all sites are attacked equally readily and, just as in an electrophile aromatic substitution such as nitration, some substituents accelerate attack at adjacent sites while other substituents retard attack. There is however a very big difference between radical substitution in aliphatic compounds and electrophilic substitution in aromatic compounds. In the latter case a particular substituent is activating and $o : p$ directing for almost every attacking agent (i.e. be it NO_2^+, SO_3, $CH_3CO^+ \cdots AlCl_4^-$ etc.). In radical transfer reactions the site most reactive to attack by one radical may be the least reactive to attack by another. This was first clearly demonstrated by the reactions of propanoic acid with alkyl radicals which attacked the α-position and with chlorine atoms which attacked the β-position preferentially.

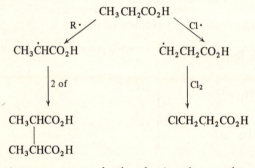

The reactions are not completely selective, the actual product ratios are $\alpha : \beta = 7.8 : 1$ for alkyl radicals and $\alpha : \beta = 0.03 : 1$ for chlorine atoms.

Remembering that different radicals show different responses to each type of substituent we will consider first unsubstituted alkanes, that is the difference in reactivity of primary, secondary and tertiary sites in aliphatic hydrocarbons. Not all primary sites are the same but quite universally the reactivity is in the order tertiary $>$ secondary $>$ primary, although the selectivity varies tremendously from one atom or radical to another. Table 8.1 which has been compiled from a wide range of sources, shows average values which have been rounded off (agreement between various sets of data is such that the table correctly shows the relative trends and orders of magnitude but is unlikely to agree exactly with any particular set of data).

The table demonstrates the huge range of reactivity for different atoms and radicals. The table also confirms that there is no simple relationship between E, the activation energy, and ΔH, the heat of reaction. Notice also that, as predicted above, the pre-exponential 'A' factors for atom reactions are on average three orders of magnitude larger than those for radicals.

Table 8.1. *The Relative Selectivities, RS_p^s, of different atoms and radicals and Arrhenius parameters for abstraction of primary hydrogen atoms*

Atom or radical	Temperature (°C)	Relative Selectivity $CH_3- \quad CH_2< \quad CH\lessgtr$			Arrhenius parameters[a] E^p (kJ mol^{-1})	$\log A^p$ (l mol^{-1} s^{-1})	ΔH (kJ mol^{-1})
F ·	25	1	1	2	1.2	11	−150
Cl ·	25	1	4	7	4.1	11	−20
tBuO ·	25	1	8	36	33	8	−17
CF$_3$ ·	150	1	8	90	35	8	−29
CH$_3$ ·	150	1	10	80	50	9	−25
Br ·	150	1	80	1600	58	12	+46
CCl$_3$ ·	150	1	80	2300	60	9	+33
I ·	150	1	1100	97000	120	12	+112

[a] Superscript p represents primary.

Table 8.2. *The Relative Selectivities, RS_p^s, of different atoms and radicals abstracting hydrogen from 1-fluorobutane in the gas phase*

Atom or radical	Temperature (°C)	FCH$_2$————CH$_2$————CH$_2$————CH$_3$ (α) (β) (γ) (δ)			
F ·	25	<0.3	0.8	1	1
Cl ·	25	0.9	2	4	1
t-BuO ·	25	7	3	8	1
CF$_3$ ·	150	2	2	9	1
CCl$_3$ ·	150	7	–	80	1
Br ·	150	10	9	80	1

The magnitude of the effect substituents have on the selectivity of free-radical transfer reactions can best be illustrated by reactions involving 1-substituted butanes. Comparison of table 8.2 (which like table 8.1 consists of rounded-off average values) with table 8.1 shows that the substituent fluorine atom has no appreciable effect on hydrogen abstraction at the γ-carbon atom or further down the chain. The β-position is deactivated to attack by all the atoms and radicals shown, but the α-position is *activated* to attack in comparison to the other terminal position (δ-position), except to attack by fluorine and chlorine atoms. These results are in marked contrast to the results obtained with 1,1,1-trifluoropentane, where the α-position is strongly de-activated to all the attacking species. The selectivities at the β-position are the same within experimental error for both molecules.

Before discussing the various effects which are responsible for this selectivity we shall look at the effect of substituents on hydrogen abstraction by chlorine atoms, since this is the most studied species. Table 8.4 shows that the substituent only affects the first two carbon atoms and that any effect further down the chain is too small to be observed. All substituents except t-butyl

Table 8.3. *The Relative Selectivities, RS_p^s, of different atoms and radicals abstracting hydrogen from 1,1,1-trifluoropentane in the gas phase*

Atom or radical	Temperature (°C)	CF_3CH_2— α	—CH_2— β	—CH_2— γ	—CH_3 δ
Cl·	25	0.03	1	4	1
CF_3·	150	0.1	2	9	1
CCl_3·	150	0.6	9	80	1
Br·	150	<1	7	80	1

Table 8.4. *The Relative Selectivities, RS_p^s, for the chlorination of 1-substituted butanes at 50°C in the gas phase*

X——————	CH_2————— (α)	CH_2————— (β)	CH_2——— (γ)	CH_3 (δ)
H—	1	4	4	1
F—	0.9	2	4	1
Cl—	0.8	2	4	1
Br—	0.4	–	4	1
CF_3—	0.04	1	4	1
FOC—	0.08	2	4	1
ClOC—	0.2	2	4	1
CH_3OOC—	0.4	2	4	1
HCOO—	–	1(.5)	4	1
CH_3COO—	0.1	2	4	1
CF_3COO—	0.2	1	4	1
N≡C—	0.2	2	4	1
CH_3O—	4	1	4	1
$(CH_3)_3C$—	3	4	5	1
C_6H_5—	7	1	(4)	1
CH_2=CH—	4	1	4	1

deactivate the β-position and all substituents except the last four also deactivate the α-position (cf. table 8.1).

We must seek to explain the substituent effects in terms of the thermochemistry, polarity and steric hindrance. We have already drawn attention to the fact that activation energies for hydrogen abstraction follow the heat of reaction but that there is in general no direct correlation between the two. When we come to consider a narrow series of very similar reactions in which polar effects are either absent or similar throughout the series then a direct correlation may be observed. Relationships of this kind are usually called Evans–Polyani equations:

$$E = \alpha \Delta H + \beta$$

(E is the activation energy, ΔH is the heat of reaction and α and β are constants; $0 < \alpha < 1$). Figure 8.2 shows the correlation for methyl and tri-

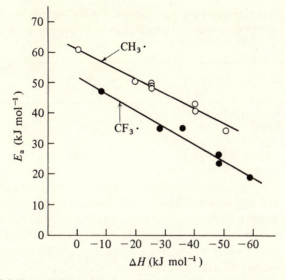

Figure 8.2. Evans–Polanyi plots for methyl and trifluoromethyl radicals reacting with alkanes.

fluoromethyl reacting with alkanes; similar correlations have been observed for other reactions of the type R—H + X· → R· + HX (where X = Cl, CF₃), and the relationship also holds for R—Cl + M → R· + MCl (where M = alkali metal atom). In the former examples R varies and X remains constant, in the latter case R remains constant but M varies.

The existence of Evans–Polyani relations, even though they are restricted to very narrow series of reactions, emphasizes the importance of the strength of the bond being broken. However table 8.2 is sufficient to show that bond strength alone is quite inadequate to explain the observed Relative Selectivities. All the radicals and atoms in table 8.2 are 'electrophilic', that is to say they are more electronegative than carbon so that the deactivation of the β-position can be attributed to the creation of a dipole in the transition state which is in opposition to the dipole of the rest of the molecule.

$$\underset{\overset{|}{R}}{\overset{\overset{H}{|}}{FCH_2-C}}-H + \cdot X \longrightarrow \underset{\overset{|}{R}}{\overset{\overset{H}{|}}{FCH_2-C}} \overset{\overset{\leftrightarrow}{}}{\cdots} H \overset{\longleftrightarrow}{\cdots} X \longrightarrow \underset{\overset{|}{R}}{\overset{\overset{H}{|}}{FCH_2-C}} \cdot + H - X$$

Much more direct evidence for the importance of polar effects comes from a comparison of the reaction of methyl and trifluoromethyl radicals with molecular hydrogen and molecular hydrogen chloride (see table 8.5). In both pairs of reactions the methyl radical is involved in an endothermic process while the trifluoromethyl radical is involved in an exothermic one. The activation energy for hydrogen abstraction from molecular hydrogen by the tri-

Table 8.5. *The activation energies and heats of reaction for the reaction of methyl and trifluoromethyl radicals with molecular hydrogen and hydrogen chloride*

	E (kJ mol^{-1})	ΔH (kJ mol^{-1})
$CH_3\cdot + H_2 \rightarrow CH_4 + H\cdot$	52	+5
$CF_3\cdot + H_2 \rightarrow CF_3H + H\cdot$	42	−12
$CH_3\cdot + HCl \rightarrow CH_4 + Cl\cdot$	10	+4
$CF_3\cdot + HCl \rightarrow CF_3H + Cl\cdot$	22	−13

fluoromethyl radical is less than that for the methyl radical as expected, but for hydrogen abstraction from hydrogen chloride it is *greater*, although the ratio of the heats of reaction remain the same. In the reactions with hydrogen chloride, the polar effect will assist the attack by methyl radicals, but oppose attack by trifluoromethyl radicals.

The Hammett equation is widely used to demonstrate the presence of polar effects in organic chemistry. The equation relates rate or equilibrium constants for processes involving a side chain on a benzene ring with the dissociation constants of the corresponding benzoic acid. The equation states;

$$\log k_{ij} - \log k_{0j} = \rho_j \sigma_i$$

where k_{ij} is a rate or equilibrium constant of reaction j when the reactant carries the substituent i, and k_{0j} is the corresponding constant for the unsubstituted reactant. The substituent constant σ_i is derived from the dissociation constant of the corresponding benzoic acid:

$$\sigma_i = \log K_i - \log K_0$$

where K_0 is the dissociation constant of benzoic acid and K_i is the dissociation constant of the appropriately substituted benzoic acid. The remaining constant ρ_i is characteristic of the reaction and in general the larger ρ_i the more important the polar effect.

The Hammett equation has been found to give reasonable correlations with a very large number of reactions, especially those which are believed to involve ionic or highly polar transition states. It was therefore of considerable interest when it was found that the relative rates of the side-chain halogenation of *meta-* and *para-*substituted toluenes gave good correlations with the Hammett equation ($\rho_{Cl}\cdot \approx -0.66$; $\rho_{Br}\cdot \approx -1.76$). Similar correlations have subsequently been established for other radical reactions and a wide variety of extended Hammett equations have been developed. The significant fact remains that hydrogen abstraction by radicals can be affected by substituents in the same way in which the substituents affect the dissociation of benzoic acids (i.e. a very polar process). Further evidence for polar effects comes from halogen abstraction and will be discussed in section 8.7.

The low reactivity of the β-position in the α-substituted butanes can thus in many cases be attributed as least in part to polar forces. When we come to compare the attack of different atoms and radicals at the α-position we find a substituent which is activating to one atom or radical may be deactivating to another. This is due to conflicting polar and bond-strength effects. A substi-

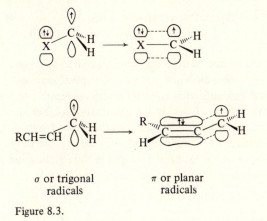

σ or trigonal π or planar
radicals   radicals

Figure 8.3.

tuent with filled p-atomic orbitals or with π-molecular orbitals can in many instances interact with an adjacent radical centre, more especially if the radical is planar. This interaction must be stabilizing since the combination of the two orbitals will lead to two new molecular orbitals, one bonding and one antibonding; but because there are only three electrons, two will go into the bonding and one into the antibonding to give overall an energy-lowering effect. Thus in the first case we discussed, namely propanoic acid, attack by alkyl radicals is directed to the α-position, because the α-radical ($CH_3 \dot{C}HCO_2 H$) is stabilized by interaction with the π-orbital of the carbonyl, thereby loosening the α-carbon–hydrogen bond. When a chlorine atom attacks, this stabilization of the incipient radical at the α-position is more than offset by the polar effect. Comparison of tables 8.3 and 8.2 clearly illustrates competition between polarity and bond strength in hydrogen abstraction. At the α-position in 1-fluorobutane the incipient radical is stabilized by interaction with the non-bonded $2p$ electrons of the fluorine and, in comparison with the other terminal position (δ), attack is facilitated except by chlorine and fluorine atoms where the opposing polar effect will be a maximum. In 1,1,1-trifluoro-pentane on the other hand the α-position is strongly deactivated to atoms and radicals. At the β-position the polar effect (similar for CF_3CH_2- and FCH_2-) is the controlling factor.

Table 8.4 shows that steric effects have little influence on hydrogen abstraction. With 2,2-dimethylhexane the α-position is more reactive than the δ-position to chlorination. The slightly enhanced attack at the δ-position in this molecular is probably due to rearrangement, involving initial attack on one of the t-butyl methyl groups followed by internal hydrogen transfer (see chapter 15).

$$(CH_3)_3C(CH_2)_3CH_3 \xrightarrow{Cl\cdot} \overset{\overset{\displaystyle CH_3}{|}}{\underset{\overset{\displaystyle |}{CH_3}}{CH_2C}} -(CH_2)_3CH_3 \longrightarrow (CH_3)_2C\overset{\overset{\displaystyle CH_2}{\diagup\diagdown}CHCH_3}{\underset{\diagdown}{\underset{\displaystyle CH_2\cdot}{}}}$$

Although steric effects do not seem to play a major role in hydrogen abstraction there is considerable evidence that the subsequent reaction between the alkyl radical and molecular chlorine is very susceptible to steric factors. The simplest example was found in the chlorination of 2-chlorobutane. All the expected isomers were obtained but the proportions of the 2,3-isomers were not equal, at 75°C the proportion of the *erythro* to the *threo* was 2 : 1. A similar example has been observed in the chlorination of

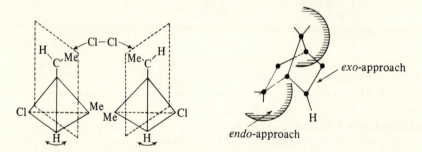

Figure 8.4. Chlorination of 2-chlorobutane at the 3-position yields 2 : 1 *erythro : threo* 2, 3-dichlorobutane; chlorination of norbornane yields 3 : 1 *exo : endo* 2-chloronorbornane.

norbornane (bicyclo[2,2,1,]heptane), where the most reactive site is the 2-position, and of the two possible isomers the *exo*-compound predominated. Similarly in the chlorination of substituted cycloalkanes the *trans* form of the 1,2-isomer always predominates over the *cis* form. In all these reactions the stereochemistry depends on the reaction of the alkyl radical with molecular chlorine. The alkyl radical is expected to be planar (sp^2-hybridized carbon, with the odd electron in a $2p$ atomic orbital). The explanation of the stereo-selectivity probably differs from reaction to reaction. In some cases (e.g. norbornane) it is simply because it is easier for the molecular halogen to approach from the *exo*-direction, and a similar argument can be extended to the 2-halobutanes. In the cycloalkanes polarity may be important. The factors which govern the preferred conformations of β-halogenoalkyl radicals are not fully understood, and the problem is further complicated by the bridging atom hypothesis (see below).

8.4. The vicinal effect, and the bridging atom hypothesis

So far we have assumed that the radical formed in the hydrogen abstraction process is sufficiently stable to remain unchanged before it is decomposed, either by abstracting an atom (usually a halogen) from another molecule or by recombining. However, attack at a position β to a substituent halogen yields a radical which can lose the substituent atom to yield an alkene; this is widely known as the vicinal effect.

$$R\dot{C}HCHClR' \rightleftharpoons RCH{=}CHR' + Cl\cdot$$

The reaction is the reverse of chlorine addition to the double bond (see also chapter 9). At room temperature the equilibrium lies well to the left, but at temperatures above $100°C$ the equilibrium shifts towards the right. If the substituent halogen is bromine then even at room temperature the equilibrium lies to the right and the gas phase chlorination of 1-bromobutane yields more 1,2-dichlorobutane than 1-bromo-2-chlorobutane:

$$BrCH_2 CH_2 C_2 H_5 \xrightarrow{Cl\cdot} BrCH_2 \dot{C}HC_2 H_5 \xrightarrow{-Br\cdot} CH_2{=}CHC_2 H_5$$
$$\downarrow Cl_2 \qquad\qquad\qquad \downarrow Cl_2$$
$$BrCH_2 CHClC_2 H_5 \qquad ClCH_2 CHClC_2 H_5$$

This is a general reaction and at higher temperatures, substituents other than halogen may be lost:

$$C_3 H_7 CH_2 OCH_3 \xrightarrow{Cl\cdot} C_3 H_7 \dot{C}HOCH_3 \longrightarrow C_3 H_7 CHO + CH_3 \cdot$$

All the results we have discussed so far have been for reactions in the gas phase. This is because in solution there are solvent effects which are only partly understood. There is however one reaction which we must discuss alongside the vicinal effect. In the liquid phase, bromination of 1-chlorobutane leads to a mixture of bromochlorobutanes very similar to those observed in the gas phase. However bromination of 1-bromobutane in the liquid phase leads almost exclusively to the formation of 1,2-dibromobutane. Two explanations have been put forward. The first explanation draws attention to the fact that hydrogen abstraction by bromine atoms is endothermic and reversible. In the gas phase the chances of re-encounter of the alkyl radical and hydrogen bromide are remote. In the liquid phase there will be many such

$$RH + Br\cdot \rightleftharpoons R\cdot + HBr$$

encounters before the radical and hydrogen bromide diffuse out of the solvent cage. At the 2-position however, the fastest process will be the unimolecular loss of a bromine atom to give but-1-ene which can then subsequently add molecular bromine. The combined effect of reversibility at all other positions and loss of a bromine followed by addition of molecular bromine would lead to apparently favoured attack at the β-position. If the reaction is carried out in the presence of DBr, deuterium is incorporated but not at a rate sufficient to be consistent with this vicinal mechanism.

The alternative explanation suggests that the substituent bromine atom bridges carbon atoms 1 and 2, and its ability to do this leads to a greatly accelerated attack at the β-position, so called 'anchimeric assistance'. Attempts to confirm the existence of a 'bridging' bromine atom have been singularly unsuccessful (see the discussion in chapter 5), nor has it been possible to confirm that the absolute rate (as distinct from the relative rate) at the β-position is abnormally fast. The problem is further complicated by the fact that HBr catalyses the formation of alkenes from bromoalkanes. The present state of knowledge leaves this problem unresolved.

8.5. Allylic halogenation

According to the concepts we developed to account for the directive effects, the position next to an alkenic double bond, the allylic position, should be very reactive. The resultant allylic radical will have the unpaired electron equally shared at the extremities of the π-orbitals extending over the three

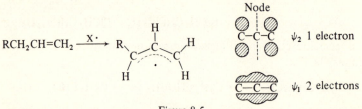

Figure 8.5.

atoms. There is however an additional problem; any radical which will abstract a hydrogen may also add to the initial double bond. We can represent a halogenation of propene as follows:

$$X\cdot + CH_3CH=CH_2 \begin{array}{c} \overset{k_2}{\nearrow} \\ \underset{k_{-2}}{} \\ \overset{k_3}{\searrow} \end{array} \begin{array}{l} CH_3\dot{C}HCH_2X \xrightarrow[k_4]{X_2} CH_3CHXCH_2X + X\cdot \\ \\ \dot{C}H_2\text{-}CH\text{-}\ddot{C}H_2(+HX) \xrightarrow[k_5]{X_2} CH_2XCH=CH_2 + X\cdot \end{array}$$

The rate of substitution to give $CH_2XCH=CH_2$ compared to the rate of addition to give CH_3CHXCH_2X is given by:

$$\frac{\text{Rate }(CH_2XCH=CH_2)}{\text{Rate }(CH_3CHXCH_2X)} = \frac{k_3}{k_2}\left(1 + \frac{k_{-2}}{k_4[X_2]}\right)$$

At low temperatures $k_2 > k_3 > k_{-2}$, but as the temperature rises both k_3 and k_{-2} will increase far more rapidly than k_2 (because the activation energies are in the order $E_{-2} > E_3 > E_2$). Thus allylic substitution is favoured by high temperatures and low molecular halogen concentration. Allylic bromination can be achieved by using N-bromosuccinimide as the source of bromine. The reaction depends on an initiator, often the presence of traces of molecular bromine. Once the reaction has started, the function of the succinimide is to furnish a constant but very low concentration of molecular bromine. The chain process is as follows:

$$Br \cdot + CH_3 CH{=}CHCO_2 Et \rightleftharpoons CH_3 CHBr\dot{C}HCO_2 Et$$

$$Br \cdot + CH_3 CH{=}CHCO_2 Et \rightleftharpoons \dot{C}H_2 CH{=}CHCO_2 Et + HBr$$

$$HBr + \begin{array}{c} CH_2CO \\ | \qquad\quad \diagdown \\ \qquad\qquad NBr \\ | \qquad\quad \diagup \\ CH_2CO \end{array} \longrightarrow Br_2 + \begin{array}{c} CH_2CO \\ | \qquad\quad \diagdown \\ \qquad\qquad NH \\ | \qquad\quad \diagup \\ CH_2CO \end{array}$$

$$\dot{C}H_2 CH{=}CHCO_2 Et + Br_2 \rightarrow BrCH_2 CH{=}CHCO_2 Et + Br \cdot$$

This reaction is much used for synthetic purposes. At one time it was believed that succinimidyl radicals were involved. It is now known that these are extremely reactive unselective species which will abstract any available hydrogen.

8.6. Solvent effects and reactions in solution

Although it is much easier to understand directive effects and elucidate reaction mechanisms in the gas phase, in practice many reactions, especially brominations, are carried out in the liquid phase. Immediately the system becomes more complicated. The chlorination of an alkane is more selective in the gas phase than in carbon tetrachloride solution. This increased selectivity is entirely due to an increase in the ratio of the pre-exponential terms, i.e. to entropy effects; in fact the activation energies are more different in solution than in the gas phase. This suggests that solvation may be playing an important role. Confirmation of this comes when we look at substituent effects in the gas and liquid phases (table 8.6). These data confirm that the substituent fluorocarbonyl group has no effect on hydrogen abstraction by chlorine atoms in the gas phase beyond the β-position. In acetonitrile solution however, a very similar substituent appears to have an effect at least as far as the ϵ-substituent. There is little doubt that the solvent facilitates charge separation in the transition state.

A more profound solvent effect is found when carbon disulphide or aromatic hydrocarbons are used as solvents in chlorination reactions and the reaction becomes very much more selective. In the case of carbon disulphide it is possible that the chlorine atom reacts with the solvent and the greatly increased selectivity is due to a quite different radical:

$$(Cl\cdot + CS_2 \rightleftharpoons \quad \underset{S}{\overset{Cl}{\diagdown}}C\text{--}S\cdot)$$

In aromatic solvents, either the chlorine atom must complex with the solvent and thereby become less reactive (but more selective), or the transition states must be solvated differently. The first explanation would require the overall rate of reaction to be slower than in a non-complexing solvent, while the second explanation would require it to be faster. Unfortunately the only experimental data refer to relative rates so that the source of the greatly increased selectivity remains unknown (table 8.7).

Table 8.6. *Relative selectivities of chlorination,* RS^x

(a) *for n-heptanoyl fluoride in the gas phase at 60°C*

FOC—CH$_2$—CH$_2$—CH$_2$—CH$_2$—CH$_2$—CH$_3$					
(α)	(β)	(γ)	(δ)	(ϵ)	(ϕ)
0.05	0.40	1	1.0	1.1	0.25

(b) *for n-heptanoyl chloride in acetonitrile solution at 52°C*

ClOC—CH$_2$—CH$_2$—CH$_2$—CH$_2$—CH$_2$—CH$_3$					
(α)	(β)	(γ)	(δ)	(ϵ)	(ϕ)
0.06	0.50	1	1.3	1.5	1.1

Table 8.7. *The relative selectivities,* RS_t^p, *in the chlorination of 2,3-dimethyl-butane in solution at 55°C*

Solvent (concentration)	RS_t^p
None	3.7
C$_6$H$_6$ (2M)	8.0
C$_6$H$_6$ (4M)	14.6
C$_6$H$_6$ (8M)	32
CH$_3$C$_6$H$_5$ (4M)	15.4
m-(CH$_3$)$_2$C$_6$H$_4$ (4M)	22.4
1,3,5(CH$_3$)$_3$C$_6$H$_3$ (4M)	25

p = primary. t = tertiary.

8.7. Abstraction of atoms other than hydrogen

Halogen atoms are readily abstracted from perhaloalkanes:

$$R\cdot + CF_3I \rightarrow RI + CF_3\cdot$$

$$R\cdot + CCl_3Br \rightarrow RBr + CCl_3\cdot$$

and these reactions are important chain propagating steps in the reactions of trifluoromethyl and trichloromethyl radicals with alkanes (hydrogen abstraction) or alkenes (addition – see chapter 9). When there are hydrogen atoms present as well, both hydrogen and halogen abstraction may be observed:

$$R \cdot + CHCl_2 Br \begin{array}{c} \nearrow RH + CCl_2 Br \cdot \\ \\ \searrow RBr + CHCl_2 \cdot \end{array}$$

Table 8.8 shows the relative rates of halogen abstraction by methyl radicals. There is an enormous range in reactivity (cf. table 8.1 which shows a similar range). However the striking results are the greatly enhanced rates of iodine abstraction from trifluoromethyl iodide and bromine from trichlorobromo-

Table 8.8. *The relative rates of halogen abstraction by methyl radicals*

$$R-X + CH_3 \cdot \xrightarrow{k_1} R \cdot + CH_3-X$$
$$C_6H_5CH_3 + CH_3 \cdot \xrightarrow{k_2} C_6H_5CH_2 \cdot + CH_4$$

R	k_1/k_2		
	X = I	X = Br	X = H
CH_3	45	0.006	0.008
C_2H_5	180	–	6
s-C_3H_7	870	–	–
t-C_4H_9	1680	–	–
CCl_3	–	7400	–
CF_3	20000	–	0.04

methane. The former reaction is particularly striking; the carbon–iodine bond in trifluoromethyl iodide is only very slightly weaker than the carbon–iodine bond in methyl iodide, and yet iodine abstraction from the fluoro compound is nearly 500 times faster. This can be explained in terms of a polar effect in the transition state:

$$\overset{\delta- \;\longleftarrow\; \delta+}{CF_3-I + CH_3 \cdot \rightarrow CF_3 \cdots I \cdots CH_3 \rightarrow CF_3 \cdot + ICH_3}$$

A similar argument applies to the relative ease of hydrogen versus chlorine abstraction from chloromethanes by trifluoromethyl radicals. The activation energy for hydrogen abstraction is on average 20 kJ mol^{-1} less than that for chlorine abstraction. These results are important because if the concept of a retarding polar effect by electronegative groups on hydrogen abstraction by chlorine atoms is correct, then interchanging the position of hydrogen and halogen atoms should reverse the effect. Evidence of this comes from the above results but even more convincingly from the reactions of trimethyltin radicals.

$$\delta+ \xrightarrow{\hspace{2cm}} \delta-$$
$$R-H + Cl\cdot \rightarrow R\cdots H\cdots Cl \rightarrow R\cdot + HCl$$

$$\delta- \xleftarrow{\hspace{2cm}} \delta+$$
$$R-Cl + (CH_3)_3Sn\cdot \rightarrow R\cdots Cl\cdots Sn(CH_3)_3 \rightarrow R\cdot + (CH_3)_3SnCl$$

Although chlorine abstraction by trimethyltin radicals is a very rapid and unselective process, chlorine abstraction from 1,1,1-trichloro-2,2,2-trifluoro-ethane (CF_3CCl_3) is two and a half times faster than chlorine abstraction from 1,1,1-trichloroethane (CH_3CCl_3). The problem is complicated by the fact that polarity and bond strength are interrelated. None the less, the CF_3 group powerfully attracts electrons while the CH_3 group weakly repels them. Halogen abstraction by trialkyltin radicals and silyl radicals is used to prepare specific radicals for further study.

8.8.   Radical transfer reactions in industrial processes

Simple atomic chlorination both in solution and in the gas phase is extensively used in the petrochemical industry. The preparation of allyl chloride from propene, as the first step in the synthesis of glycerol has already been referred to (see p. 76).

$$Cl\cdot + CH_3CH{=}CH_2 \nearrow CH_3\overset{\cdot}{C}HCH_2Cl$$
$$\searrow HCl + CH_2{-}CH{-}CH_2 \xrightarrow{Cl_2} ClCH_2CH{=}CH_3$$

Chloromethanes are prepared by the direct chlorination of methane (carbon tetrachloride is also made by the chlorination of carbon disulphide), ethyl chloride by the direct chlorination of ethane, and mixtures of alkanes (from petroleum) are monochlorinated to be used as cutting oils and plasticizer extenders. Fluorthene ($CF_3CHClBr$), an important anaesthetic, is prepared by the photochemical bromination of the chloro-2,2,2-trifluoroethane.

Industrial chlorination may be thermal or photochemical, but chlorine is expensive, requiring electric power in its manufacture. One alternative is 'oxychlorination'; this process depends on the exothermic nature of the reaction:

$$2HCl + 2RH + O_2 \rightarrow 2H_2O + 2RCl \qquad \Delta H = 700 \text{ kJ mol}^{-1}$$

The exact mechanism of the reaction is unknown, but the process requires very high temperatures, and almost certainly hydroxy radicals formed by the decomposition of peroxides are important.

Perhaps the most interesting application is the photochemical nitrosation of cyclohexane to yield cyclohexanone oxime, which is converted into caprolactam and so into nylon 6:

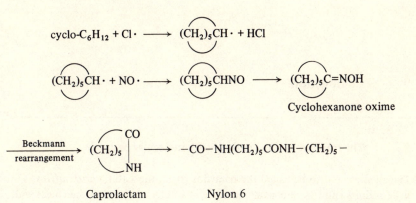

$$\text{NOCl} \xrightarrow{h\nu} \text{NO} \cdot + \text{Cl} \cdot$$

$$\text{cyclo-C}_6\text{H}_{12} + \text{Cl} \cdot \longrightarrow (\text{CH}_2)_5\text{CH} \cdot + \text{HCl}$$

$$(\text{CH}_2)_5\text{CH} \cdot + \text{NO} \cdot \longrightarrow (\text{CH}_2)_5\text{CHNO} \longrightarrow (\text{CH}_2)_5\text{C=NOH}$$

Cyclohexanone oxime

$$\xrightarrow[\text{rearrangement}]{\text{Beckmann}} (\text{CH}_2)_5 \begin{array}{c} \text{CO} \\ | \\ \text{NH} \end{array} \longrightarrow -\text{CO}-\text{NH}(\text{CH}_2)_5\text{CONH}-(\text{CH}_2)_5-$$

Caprolactam   Nylon 6

9 Combustion

9.1. The reaction of alkyl radicals with molecular oxygen

The alkanes used to be called the paraffins (from the Latin, *parum affins*, little affinity) and were considered to be characterized by their inertness and failure to react with common reagents. It is true that they show little reactivity with ionic reagents like sulphuric acid or sodium hydroxide but as we have seen they react rapidly with atoms and radicals, and indeed the internal combustion engine depends on their high reactivity towards radical reagents.

In chapter 1 we saw that molecular oxygen in the ground state is a 'triplet', that is a kind of biradical. This means that just as radicals combine with each other so they combine with molecular oxygen extremely rapidly. At normal temperatures molecular oxygen will not abstract a hydrogen atom from an alkane, but at the highly elevated temperatures encountered in flames this process becomes possible. At low temperatures, the alkylperoxy radical formed by the combination of an alkyl radical and molecular oxygen is unreactive and will not abstract hydrogen from an unactivated carbon hydrogen bond in the gas phase. In solution peroxy radicals will abstract activated hydrogen atoms and this is important in autoxidation (see chapter 14).

$$R \cdot + O_2 \rightarrow RO_2 \cdot \tag{9.1}$$

$$RO_2 \cdot + RH \rightarrow RO_2 H + R \cdot \tag{9.2}$$

$$\text{(below } 150^{\circ}C)$$

The primary fate of the alkylperoxy radicals at room temperature is disproportionation.

$$2RO_2 \cdot \rightarrow 2RO \cdot + O_2 \tag{9.3}$$

Since this is a radical–radical reaction and therefore its rate depends on the square of a low concentration, it means that at ambient temperature a trace of oxygen can act as a powerful inhibitor of chain reactions. In order to understand the explosive nature of many reactions involving oxygen we have to be familiar with the concept of *branching chains*.

9.2.   Branching-chain reactions

In a normal chain reaction one reactive centre, i.e. a radical, propagates another reactive centre so that the total number of reactive centres (radicals) remains unchanged and depends only on the rates of initiation and termination. Branching occurs when one reactive centre produces two (or more) reactive centres in a propagating step. The most common branching is *linear branching* in which the rate of production of new reactive centres is proportional to the first power of the concentration of those already present.

The simplest linear branching is *thermal branching*, and it occurs when the products of a highly exothermic step are sufficiently excited to promote the dissociation of a reactant molecule. Thus fluorination of an alkane can become uncontrolled if no inert gas is provided as a diluent.

$$F_2 \rightleftharpoons 2F \cdot \tag{9.4}$$

$$F \cdot + RH \rightarrow HF^* + R \cdot \tag{9.5}$$

$$R \cdot + F_2 \rightarrow RF^{**} + F \cdot \tag{9.6}$$

*thermally excited   **thermally highly excited

The simple fluorination sequence involves the chain propagating steps (9.5) and (9.6). However the thermally excited alkyl fluoride (RF^{**}) can transfer some of its excess energy to molecular fluorine molecules, with the result that the rate of initiation is increased:

$$RF^{**} + F_2 \rightarrow RF + 2F \cdot \tag{9.7}$$

Normal branching comprises the occasional production of two or more centres in a propagation step in which one centre is destroyed. Such reactions are usually endothermic and have a considerable energy of activation (sometimes referred to as the *energy of branching*). The reaction between oxygen and hydrogen is believed to include such steps:

$$H_2 + O_2 \rightarrow 2HO \cdot \qquad\qquad\qquad\qquad\qquad\quad (9.8)$$
$$\left. \begin{array}{c} \end{array} \right\} \textit{initiation}$$
$$H_2 + M \rightarrow 2H \cdot + M \qquad\qquad\qquad\qquad\qquad (9.9)$$

$$HO \cdot + H_2 \rightarrow H \cdot + H_2O \qquad \textit{propagation} \qquad (9.10)$$

$$H \cdot + O_2 \rightarrow HO \cdot + O \cdot \qquad\qquad\qquad\qquad\quad (9.11)$$
$$\left. \begin{array}{c} \end{array} \right\} \textit{branching}$$
$$O \cdot + H_2 \rightarrow H \cdot + HO \cdot \qquad\qquad\qquad\qquad\quad (9.12)$$

$$H \cdot \rightarrow Wall \qquad\qquad\qquad\qquad\qquad\qquad\qquad (9.13)$$
$$\left. \begin{array}{c} \end{array} \right\} \textit{termination}$$
$$H \cdot + O_2 + M \rightarrow HO_2 \cdot + M \qquad\qquad\qquad\qquad (9.14)$$

Both (9.10) and (9.11) have appreciable activation energies (75 kJ mol^{-1} and 25 kJ mol^{-1} respectively).

Continuous branching occurs when the main propagation step is also a branching reaction e.g.

$$P_4O_n \cdot + O_2 \rightarrow P_4O_{n+1}^{\cdot} + O \cdot \qquad (9.15)$$

This is the type of process involved in nuclear-chain reactions on which the development of atomic power depends.

Degenerate branching occurs when the primary chain is unbranched but additional reactive centres are produced by side reactions of the products of the primary chain. We have already drawn attention to the low reactivity of alkylperoxy radicals. If the temperature is raised sufficiently the alkylperoxy radicals will abstract hydrogen (especially from reactive sites as in aldehydes).

$$R \cdot + O_2 \rightarrow RO_2 \cdot \qquad \qquad (9.1)$$
propagation
$$RO_2 \cdot + RH \rightarrow RO_2H + R \cdot \qquad (9.2)$$

However at these temperatures the hydroperoxide is unstable and will undergo unimolecular decomposition to yield an alkoxy radical and a hydroxy radical (i.e. branching).

$$RO_2H \rightarrow RO \cdot + HO \cdot \qquad \textit{branching} \qquad (9.16)$$

9.3. Explosions

The steady-state treatment is clearly inapplicable to branching chains and we must derive a time-dependent equation for the total radical concentration $[r]$. The rate of formation of radical centres depends on the rate of initiation, (i) and the rate of branching (b) which depends on $[r]$. Since there is no net production or removal of radicals in the non-branching propagation steps we need only consider in addition the termination step (f) which also depends on $[r]$. We can thus write the following expression for the rate of formation of radical centres:

$$\frac{d[r]}{dt} = i + k_b[r] - k_f[r]$$

Writing ϕ for $(k_b - k_f)$ we have

$$\frac{d[r]}{dt} = i + \phi[r]$$

which on integration gives

$$[r] = \frac{i \times \exp(t) - 1}{\phi}$$

If ϕ is positive, i.e. if branching occurs more rapidly than termination, the radical centres accumulate and generate more chain centres, and this leads to an exponential growth, i.e. an *explosion*! Thus for the hydrogen/oxygen reaction each initial hydrogen atom generates a further 10^{16} hydrogen atoms per second at about $430°C$ and an oxygen pressure of 11 kPa. The reaction is not explosive under all conditions. Figure 9.1 shows explosion limits for a stoichio-

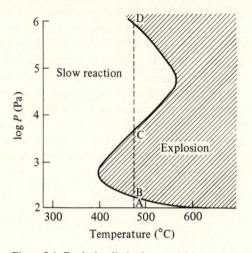

Figure 9.1. Explosion limits for a stoichiometric hydrogen–oxygen mixture.

metric hydrogen–oxygen mixture. At $480°C$ and very low pressures (A) the branching reactions are slow and the chain centres diffuse to the walls: linear termination (9.13) is faster than branching (i.e. ϕ is negative) and the reaction is slow. As the pressure increases (the temperature constant) the extent of branching increases until finally ϕ becomes positive and an explosion occurs (B). A further substantial increase in pressure (still at the same temperature) means that termolecular termination (9.14) becomes more important until it reaches a stage where it exceeds branching (ϕ is negative) and we enter another region of slow reaction (C). If we increase the pressure still further the unreactive $HO_2 \cdot$ radical starts to participate (9.17) and (9.18) since it can no longer diffuse to the walls.

$$HO_2 \cdot + H_2 \rightarrow H \cdot + H_2O_2 \qquad (9.17)$$
$$HO_2 \cdot + H_2O \rightarrow HO \cdot + H_2O_2 \qquad (9.18)$$

very slow propagation

These reactions regenerate the reactive radicals $H \cdot$ and $HO \cdot$ which take part in exothermic reactions (9.10) and (9.11); heat is generated faster than it can diffuse to the walls and in addition we have increasing concentrations of the thermally unstable peroxide so a *thermal explosion* occurs (D). We have discussed the hydrogen/oxygen reaction but very similar mechanisms apply for alkane/oxygen explosions.

9.4. Slow oxidation of alkanes

Although the explosive nature of the reaction between oxygen and the alkanes is well known, the non-explosive reaction is of considerable chemical interest. The explosive reaction usually yields carbon dioxide and water as the ultimate products, but the slow reaction produces a wide variety of alcohols, aldehydes, ketones and alkenes. The range of reactivity of different organic molecules is very great. Under conditions where acetaldehyde is oxidized at an appreciable rate at $80°C$, methane requires a temperature in excess of $400°C$ before it reacts. Just as the explosion limit varies with pressure in an irregular way, so the rate of reaction passes through a maximum with increasing temperature, usually showing a negative temperature coefficient for about $50°C$ before increasing rapidly with a further increase in temperature. The reaction products also change. In the low temperature region the products include the carbon oxides, alcohols, aldehydes and ketones. At higher temperatures alkenes become increasingly important products. The alkene-forming reactions may proceed by cyclic transition states.

$$RCH_2CH_2CH_2\cdot + O_2 \rightarrow RCH_2CH_2CH_2O_2\cdot \tag{9.19}$$

$$\tag{9.20}$$

However, because of the number of reaction sequences occurring at the same time it has not been possible to elucidate the full reaction mechanisms. Slow oxidation of butane yields among other products but-1-ene, *cis-* and *trans-*but-2-enes, butadiene, 1,2-epoxybutane, *cis-* and *trans-*2,3-epoxybutanes, 2-methyloxetan, tetrahydrofuran, propene, acetone, prop-2-en-1-ol, ethylene, acetaldehyde, methane and formaldehyde.

The negative temperature coefficient probably represents the onset of hydrogen abstraction by alkylperoxy radicals. At low temperatures these radicals principally take part in radical–radical reactions to yield more reactive alkoxy radicals.

$$RO_2\cdot + RO_2\cdot \rightarrow 2RO\cdot + O_2 \tag{9.3}$$

When the temperature rises they start to abstract hydrogen.

$$RO_2\cdot + RH \xrightarrow{\text{slow}} RO_2H + R\cdot \tag{9.2}$$

$$R\cdot + O_2 \longrightarrow RO_2\cdot \tag{9.1}$$

This leads to the chain-propagating step (9.2) which is slow compared with the rapid hydrogen abstraction (9.21) by the alkoxy radical.

$$RH + RO \cdot \xrightarrow{\text{fast}} ROH + R \cdot \qquad (9.21)$$

When the temperature rises still further the hydroperoxide dissociates to yield reactive radicals and we have degenerate branching, which will lead to an explosion.

$$RO_2H \rightarrow RO \cdot + HO \cdot \qquad (9.16)$$

Flames represent controlled explosions in which the fuels (oxygen and an alkane for example) are being supplied as fast as they react, and the products of the reaction are free to escape. Temperatures in the reacting zone can be very high; so high, in fact that ionic as well as radical processes are occurring. Further discussion of flames is outside the scope of this book.

10 Radical addition reactions

10.1. Characteristic features of radical addition reactions

Virtually all carbon-centred radicals and a great variety of heteroradicals and atoms will add to molecules containing unsaturated centres. Alkenes, alkynes, carbonyl compounds and azo-compounds are all effective partners in this reaction. Radical addition to aromatic compounds occurs with equal facility, but the overall result of the process is usually substitution, and these reactions are considered separately in chapter 12, p. 123.

When a radical adds to an alkene the double bond of the alkene is broken and a new single bond is formed between one of the alkene carbons and the incoming radical.

$$R \cdot + CHX{=}CHY \rightarrow R{-}CHX{-}CHY \cdot$$

The reaction leads to the formation of a new adduct radical, and the final products depend on the subsequent fate of this radical. Some of the more common reactions of the adduct radicals are illustrated below in the scheme:

$$
R{-}CHX{-}CHY \cdot
\begin{cases}
\xrightarrow{} R \cdot + CHX{=}CHY & \text{(i)} \\
\xrightarrow{} X \cdot + RCH{=}CHY & \text{(ii)} \\
\xrightarrow{RZ} R{-}CHX{-}CHYZ + R \cdot & \text{(iii)} \\
\xrightarrow{CHX{=}CHY} R{-}CHX{-}CHY{-}CHX{-}CHY \cdot & \text{(iv)} \\
\xrightarrow{R' \cdot} R{-}CHX{-}CHY{-}R' & \text{(v)}
\end{cases}
$$

The *fragmentation* reaction (i) is simply the reverse of the addition step and, since this is usually exothermic, reaction (i) is only important at high temperatures for alkyl radicals. However, with bromine or iodine atoms, thiyl radicals, and many other heteroradicals the reverse reaction is important even at room temperature. Loss of another group from the adduct radical (ii) can be a very facile process when the alkene contains halogen atoms or other weakly bound groups. The fragmentation reactions (i) and (ii) are treated in detail below (p. 101).

Radical *transfer* (iii) with an added transfer agent RZ, which may be the radical source, leads to a chain reaction in which the initial radical $R \cdot$ is regenerated and a 1 : 1-adduct of the alkene and the transfer agent is produced. A wide variety of organic molecules are suitable transfer agents, and this process is useful for synthesis of the 1 : 1-adducts which are formed in good yields if excess of RZ is used. For example, alkyl halides such as bromotrichloromethane will add across the double bond of alkenes in a chain reaction which can be initiated photochemically or with peroxides:

$$CCl_3 Br \rightarrow CCl_3 \cdot + Br \cdot \qquad \textit{initiation}$$

$$CCl_3 \cdot + RCH = CH_2 \rightarrow CCl_3 CH_2 \dot{C}HR$$

$$CCl_3 CH_2 \dot{C}HR + CCl_3 Br \rightarrow CCl_3 CH_2 CHBrR + CCl_3 \cdot$$

Telomerization (iv) may also occur when the adduct radical adds to another alkene unit and further telomer radicals may be formed by successive additions of this kind. The reaction products in this case consist of a series of telomers $R(CHXCHY)_n Z$, where n usually ranges from 1 to 10. Telomerization is favoured when excess alkene is used, and when the adduct radicals are more prone to add to the alkene than transfer with RZ. Fluorinated alkenes such as tetrafluoroethylene telomerize particularly readily:

$$R(CF_2 CF_2)_{n-1}^{\cdot} + CF_2 = CF_2 \rightarrow R(CF_2 CF_2)_n^{\cdot}$$

$$R(CF_2 CF_2)_n^{\cdot} + RZ \rightarrow R(CF_2 CF_2)_n Z + R \cdot$$

In the extreme case where the addition steps are very rapid, or very high concentrations of alkene are used, *polymerization* will occur in which hundreds or thousands of alkene units are incorporated into a macromolecular polymer radical before the reaction stops. Polymerization reactions are of vital industrial importance and are considered separately in chapter 11.

Combination (v) or disproportionation of the adduct radical with other radicals in the system will usually be important as a chain-termination step. However, products of radical coupling reactions are minor except when the radical concentration is high, or when the kinetic chain length is short. High radical concentrations are often generated in the neighbourhood of the electrodes in electrochemical reactions. For instance, electrolysis of alkanoic acids RCOOH in the presence of alkenes $CH_2 = CHX$ gives predominantly dimeric products $RCH_2 CHXCHXCH_2 R$.

Radicals differ greatly in their ability to add to alkenes. Atoms such as $H \cdot$ and $Cl \cdot$ add more rapidly than polyatomic radicals because they have no rotational or vibrational degrees of freedom to lose in the process. This means that the entropy factor is more favourable for atom addition, and thus the pre-exponential factors in the rate constants for addition are higher than for polyatomic radicals. Some absolute rate constants and Arrhenius parameters for addition of alkyl and haloalkyl radicals to ethylene are shown in table 10.1.

Table 10.1. *The rate constants at 164°C and Arrhenius parameters for the addition of alkyl and haloalkyl radicals to ethylene*

Radical	$\log A$ $(1\,mol^{-1}\,s^{-1})$	E $(kJ\,mol^{-1})$	k $(1\,mol^{-1}\,s^{-1})$
$CH_3 \cdot$	8.5	32.3	4.6×10^4
$C_2H_5 \cdot$	8.2	30.6	3.5×10^4
$CH_2F \cdot$	7.6	18.1	2.8×10^5
$CCl_3 \cdot$	7.8	26.4	4.5×10^4
$CF_2Br \cdot$	8.0	13.0	2.8×10^6
$CF_3 \cdot$	8.0	12.2	3.5×10^6

The rate constants vary by two orders of magnitude, from those for the slow alkyl radicals to those for the much faster trifluoromethyl radicals. In general electrophilic radicals, i.e. those with electronegative substituents adjacent to the radical centre, appear to be the most reactive. The marked difference in reactivity of the various radicals is due almost entirely to changes in the activation energy terms, and the A factors vary by less than a power of ten for the whole range of radicals.

The rate of addition depends on the structure of the alkene as well as on the nature of the attacking radical. Trifluoromethyl radicals are always more reactive than methyl radicals, but while $CF_3 \cdot$ radicals add to ethylene much more rapidly than to tetrafluoroethylene, $CH_3 \cdot$ radicals add more rapidly to tetrafluoroethylene than to ethylene. This indicates the importance of polar effects on the rate of addition; steric effects are also important (see below).

10.2. Orientation of radical addition

In the presence of an unsymmetrical alkene two adduct radicals can be formed from addition of the primary radical to each end of the double bond:

$$R \cdot + CH_2 = CHX \quad \overset{k_a}{\nearrow} \quad RCH_2CHX \cdot$$
$$\underset{k_\beta}{\searrow} \quad RCHXCH_2 \cdot$$

$$RCH_2CHX \cdot + RZ \longrightarrow RCH_2CHXZ + R \cdot$$

$$RCHXCH_2 \cdot + RZ \longrightarrow RCHXCH_2Z + R \cdot$$

Two isomeric 1 : 1-adducts are produced when these adduct radicals abstract Z· from the transfer agent RZ. In practice radicals seldom add exclusively to one end, but usually give a mixture of both possible products although one is generally formed in substantial excess. Monosubstituted alkenes $XCH=CH_2$ and 1,1-disubstituted alkenes $XYC=CH_2$ always give predominantly the adduct from addition to the unsubstituted end, irrespective of the nature of the radical or the substituents X and Y. For 1,2-disubstituted, and trisubstituted alkenes the main adduct usually, but not always, comes from addition to the least substituted end.

For a simple system in which the addition steps are not reversible, and adduct radicals are not lost in other processes like telomerization or combination, the ratio of the concentrations of the two adducts is equal to the ratio of the rate constants for addition to the two ends of the alkene. The *Orientation Ratio* **OR** is defined as the ratio of the rate constant for addition to the most substituted end (β) to the rate constant for addition to the least substituted end (α) *i.e.*:

$$\mathbf{OR} = k_\beta/k_\alpha = \text{Rate } (R \, CHXCH_2 \, Z)/\text{Rate } (R \, CH_2 \, CHXZ)$$

The orientation ratios for addition of trifluoromethyl and trichloromethyl radicals to a variety of alkenes are given in table 10.2. The table shows a wide

Table 10.2. *Orientation ratios for the addition of trifluoromethyl radicals (liquid phase) and trichloromethyl radicals (gas phase) to chloroethylenes and fluoropropenes*

Olefin $\alpha-\beta$	$CF_3\cdot$ $\alpha:\beta$	$CCl_3\cdot$ $\alpha:\beta$
$CH_2=CHCl$	$1:<0.01^{(a)}$	$1:<0.01^{(a)}$
$CHCl=CF_2$	$1:11.5$	$1:25$
$CHCl=CCl_2$	–	$1:0.03$
$CH_2=CHCH_3$	$1:0.12 \ (1:0.08)$	$1:0.07$
$CH_2=CFCH_3$	–	$1:0.007$
$CH_2=CHCF_3$	$1:<0.01^{(a)}$	$1:<0.01^{(a)}$
$CHF=CHCF_3$	$1:0.33$	–
$CF_2=CHCH_3$	$1:>50^{(b)}$	–
$CF_2=CHCF_3$	$1:1.49$	–
$CF_2=CFCF_3$	$1:0.25 \ (1:0.27)$	$1:<0.1^{(a)}$

[a] Only the adduct from addition to the α-end of the alkene was detected.
[b] Only the adduct from addition to the β-end of the alkene was detected.

range in the proportions of the two adducts depending on the substituents in the alkene and on the type of radical involved. For vinyl chloride only the adduct from addition to the CH_2- end is detected, but with others comparable amounts of both adducts are observed. Both radicals add preferentially to the more substituted end of 1,1-difluorochloroethylene.

The orientation of electrophilic addition to any unsymmetric alkene is usually rationalized in terms of the extent of resonance stabilization in the adduct carbonium ion. Addition occurs so as to give the more stable of the two possible carbonium ions:

$$E^+ + \ddot{X}-CH{=}CH_2 \rightarrow \ddot{X}-\overset{+}{C}H-CH_2E \leftrightarrow \overset{+}{X}{=}CH-CH_2E$$

It is very tempting to adopt a similar argument for radical addition:

$$R\cdot + \ddot{X}-CH{=}CH_2 \rightarrow \ddot{X}-\dot{C}H-CH_2R \leftrightarrow \overset{+\cdot}{X}-\overset{\cdot\cdot}{\underset{\cdot\cdot}{C}}H-CH_2R$$

It will be seen that resonance delocalizes the charge in the adduct carbonium ion formed during electrophilic addition, whereas in the adduct radical formed in the homolytic reaction resonance creates charge separation where there was none. This means that, according to resonance theory, the dipolar structure is unlikely to contribute extensively to the ground state of the adduct radical. We shall see below that although resonance stabilization may play a small part in controlling the orientation of radical addition it is only one contributing factor and often not the deciding one.

In table 10.3 the orientation ratios for addition of a variety of radicals to

Table 10.3. *Orientation ratios ($\alpha : \beta$) for the addition of alkyl radicals to vinyl fluoride, 1,1-difluoroethylene and trifluoroethylene at 150°C*

Radical	α β $CH_2{=}CHF$	α β $CH_2{=}CF_2$	α β $CHF{=}CF_2$
$CF_3\cdot$	1 : 0.09	1 : 0.03	1 : 0.50
$CHF_2\cdot$	1 : 0.19	1 : 0.15	1 : 0.95
$CH_2F\cdot$	1 : 0.30	1 : 0.44	1 : 2.04
$CH_3\cdot$	1 : 0.20	–	1 : 2.10
$CCl_3\cdot$	1 : 0.07	1 : 0.01	1 : 0.29
$CH_2Cl\cdot$	1 : 0.18	1 : 0.14	1 : 1.03
$CBr_3\cdot$	1 : 0.04	–	1 : 0.24
$CHBr_2\cdot$	1 : 0.06	–	1 : 0.31
$CF_3\cdot$	1 : 0.09	1 : 0.03	1 : 0.50
$CF_2Br\cdot$	1 : 0.09	1 : 0.03	1 : 0.47
$CFBr_2\cdot$	1 : 0.08	1 : 0.02	1 : 0.37
$CBr_3\cdot$	1 : 0.04	–	1 : 0.24
$CF_3\cdot$	1 : 0.09	1 : 0.03	1 : 0.50
$CF_3CF_2\cdot$	1 : 0.05	1 : 0.01	1 : 0.29
$(CF_3)_2CF\cdot$	1 : 0.02	1 : 0.001	1 : 0.06
$(CF_3)_3C\cdot$	1 : 0.005	–	–
$CF_3CF_2\cdot$	1 : 0.054	1 : 0.011	1 : 0.29
$CF_3(CF_2)_2\cdot$	1 : 0.050	1 : 0.009	1 : 0.25
$CF_3(CF_2)_3\cdot$	1 : 0.050	1 : 0.007	1 : 0.24
$CF_3(CF_2)_6\cdot$	1 : 0.049	1 : 0.007	1 : 0.23
$CF_3(CF_2)_7\cdot$	1 : 0.043	1 : 0.006	1 : 0.22

vinyl fluoride, 1,1-difluoroethylene, and trifluoroethylene are shown. All the radicals add preferentially to the CH_2- end of both vinyl fluoride and 1,1-difluoroethylene, although the orientation ratio varies considerably from radical to radical. However, with trifluoroethylene the most halogenated radicals add preferentially to the $CHF-$ end of the molecule while methyl and monohalomethyl radicals add preferentially to the more substituted CF_2- end. This observation and the strong dependence of the orientation ratios on the nature of the attacking radical, shows that delocalization of the unpaired electron in the adduct radical, either by resonance or by hyperconjugation, is not the *prime* factor governing the orientation of addition to alkenes. Although there is no reversal of orientation, a similar trend is to be noted for the other two alkenes: the proportion of attack at the more substituted end of the alkene decreases as the radical carries more halogen atoms. In the series $CX_3\cdot$, $CHX_2\cdot$, $CH_2X\cdot$, $CH_3\cdot$, where X represents a halogen atom, the electronegativity steadily decreases and the orientation ratios suggest that polar forces play a significant role in determining the orientation of radical addition to alkenes.

The same trends are observed in the addition of a series of phosphinyl radicals of decreasing electronegativity $((CF_3)_2P\cdot, H_2P\cdot, Me_2P\cdot)$ to the fluoroethylenes, as table 10.4 shows. This constitutes further evidence of the

Table 10.4. *Orientation ratios for the addition of phosphinyl radicals to fluoroethylenes*

Radical	α β $CH_2=CHF$	α β $CH_2=CF_2$	α β $CHF=CF_2$
$Me_2P\cdot$	1 : 0.09	1 : 0.39	1 : 1.08
$H_2P\cdot$	–	1 : 0.00[a]	1 : 0.19
$(CF_3)_2P\cdot$	1 : 0.00[a]	1 : 0.00[a]	1 : 0.02

[a] Only the adduct from addition to the α-end of the olefin was detected.

importance of polar effects in radical addition reactions. The logarithms of the orientation ratios show an approximately linear correlation with the Hammett σ-constants of the attacking radicals, which also indicates the importance of polar effects.

In the series $CF_3\cdot$, $CF_2Br\cdot$, $CFBr_2\cdot$, $CBr_3\cdot$, the change in electronegativity is small. By analogy with the fluoromethyl series a small increase in the orientation ratio along the series would be expected, but in fact it decreases (see table 10.3). This result suggests that classical steric repulsion due to the bulk of the bromine atoms is important. Stronger evidence for classical steric hindrance comes from the series $CF_3\cdot$, $CF_3CF_2\cdot$, $(CF_3)_2CF\cdot$, $(CF_3)_3C\cdot$, which shows a large increase in the orientation ratio along the series of radicals for all three alkenes. In contrast, the orientation ratios for the straight-chain radicals $C_nF_{2n+1}\cdot$ show almost negligible change from $n = 2$ to $n = 8$. The anomalous addition of $CF_3\cdot$, and $CCl_3\cdot$ radicals to the more substituted end of 1,1-difluorochloroethylene can also be attributed to steric repulsion by the large chlorine atom to approach of the radicals at the CHCl— end.

10.3. Carbon–carbon bond formation by radical addition

The addition of carbon-centred radicals to unsaturated compounds is a useful method for the synthesis of new carbon–carbon bonds. The basic mechanism for formation of a 1 : 1-adduct between the molecule RZ and the alkene E involves a chain reaction:

$$RZ + In\cdot \xrightarrow{k_d} R\cdot + InZ$$

$$R\cdot + E \xrightarrow{k_p} RE\cdot$$

$$RE\cdot + RZ \xrightarrow{k_a} REZ + R\cdot$$

$$R\cdot + R\cdot \xrightarrow{k_c} R_2$$

Peroxides, azo-compounds and ultraviolet light are the most commonly used initiators. In the presence of excess RZ the main chain-termination reaction will involve combination of two primary radicals. Since the addition step is the slow, rate-controlling reaction, the rate of formation of the adduct is given by:

$$\frac{d[REZ]}{dt} = k_p[R\cdot][E],$$

and the rate of dimer formation by:

$$\frac{d[R_2]}{dt} = k_c[R\cdot]^2 = \tfrac{1}{2}V_i$$

where V_i is the rate of initiation. Hence, substituting for the radical concentration:

$$\frac{d[REZ]}{dt} = k_p[E]\left(\frac{d[R_2]/dt}{k_c}\right)^{\frac{1}{2}} = k_p[E](V_i/2k_c)^{\frac{1}{2}}$$

To obtain good yields of the 1 : 1-adduct the alkene must be very reactive, i.e. k_p must be large, and the chains must be as long as possible to minimize the loss of reactant in termination reactions and to increase the efficiency of conversion. Low radical concentrations favour long-chain processes, and this can be achieved by using small initiator concentrations and slow rates of decomposition (i.e. low temperatures), or, in photochemical reactions, low light intensities. However, low radical concentrations decrease the rate of adduct formation, thus lengthening the time for complete conversion of the reactants. An acceptable compromise must be worked out for each individual situation.

Formation of 2 : 1-, 3 : 1-, and other telomers can be important side reactions which reduce the yield of 1 : 1-adduct. These undesirable by-products can usually be suppressed or minimized by working with a high ratio of RZ to alkenes, or by increasing the temperature of the reaction. When the alkene contains allylic hydrogen atoms, allylic abstraction by the primary radicals can be a major complication.

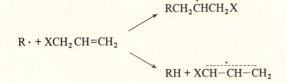

$$R \cdot + XCH_2 CH = CH_2 \begin{cases} RCH_2 \overset{\cdot}{C}HCH_2 X \\ \\ RH + XCH - \overset{\cdot}{CH} - CH_2 \end{cases}$$

It is usually more important for cyclic and non-terminal alkenes than for 1-alkenes, and it can often be reduced to acceptable proportions by working at lower temperatures.

Polyhaloalkanes such as CCl_4 and $CCl_3 Br$ are particularly useful for the synthesis of 1 : 1-adducts which may be important in their own right or as intermediates in the productions of other molecules, as in the following example:

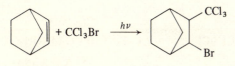

$$\text{Norbornene} + CCl_3 Br \xrightarrow{h\nu}$$

Norbornene

The trihalomethyl group in the adduct can be hydrolysed under mild conditions to give the corresponding carboxylic acid, and the presence of the halogen means that the adducts can be easily converted into ethers, alcohols, amines or organometallic reagents. In the transfer step the weakest carbon–halogen bond is broken, so that for mixed polyhaloalkanes the order of halogen transfer is: $I > Br > Cl$, e.g.:

$$CFCl_2 Br + RCH = CH_2 \rightarrow RCHBrCH_2 CFCl_2$$

The best-behaved haloalkanes are those with at least three halogen atoms. For polyhalomethanes containing one or more hydrogen atoms of the type $CHX_2 Br$ and $CHX_2 I$, hydrogen transfer may be competitive with bromine or iodine transfer, and a mixture of adducts will result. For example, in the photochemical addition of $CHCl_2 Br$ to alkenes, adducts from both $CHCl_2 \cdot$ radicals and $CCl_2 Br \cdot$ radicals were observed in approximately equal proportions. When the molecule contains two or more bromine or iodine atoms, photochemical initiation gives a substantial proportion of carbene accompanying the normal radical. The carbenes undergo cyclo-addition with the alkene and the resulting cyclopropane derivatives may be important by-products.

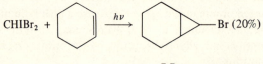

7-Bromonorcarane

The radical produced from a haloalkane is generally electrophilic in behaviour so that addition occurs best with electron-rich alkenes containing alkyl and alkoxy substituents. Excellent yields are also obtained in the addition of polyfluoroalkyl iodides to fluorinated alkenes because the iodine transfer step with the adduct radical is particularly easy:

$$CF_3CF{=}CF_2 + CF_3I \xrightarrow{h\nu} CF_3CFICF_2CF_3 \qquad (94\%)$$

Aldehydes, ketones and alcohols also add to alkenes by essentially the same process. The overall result of the reactions is shown by the following examples:

$$C_3H_7{}^nCHO + CH_2{=}CH(CH_2)_2COCH_3 \xrightarrow{\text{peroxide}} C_3H_7{}^nCO(CH_2)_4COCH_3$$

$$(70\%)$$

$$C_2H_5OH + C_3F_7{}^nCF{=}CF_2 \xrightarrow{\text{peroxide}} C_3F_7{}^nCHFCF_2CH(OH)CH_3$$

$$(70\%)$$

The addition reaction has also proved particularly useful for the alkylation of saturated heterocycles using 1-alkenes. Using di-t-butyl peroxide, the t-butoxy radical first abstracts hydrogen from the carbon atom adjacent to the heteroatom: the overall reaction is as shown.

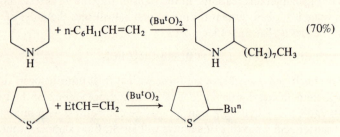

Alkylation always occurs α to the heteroatom, but this is usually accompanied by β-alkylation for six-membered rings.

Carbon-centred radicals also add to alkynes, but not so readily as to alkenes. The products consist of a mixture of *cis*- and *trans*-adducts, usually in comparable proportions. These adducts are themselves unsaturated, and it is common for further addition to take place giving products incorporating two molecules of the addend. This occurs in the peroxide-initiated addition of acetaldehyde to phenylacetylene, although it is possible to isolate the intermediate *cis*- and *trans*-4-phenyl-3-buten-2-ones (1) if photochemical initiation is used.

$$PhC{\equiv}CH$$
$$+ \quad \xrightarrow{h\nu} \quad CH_3COCH{=}CHPh \quad \xrightarrow[\text{peroxide}]{CH_3CHO}$$
$$CH_3CHO$$

$$CH_3COCH(Ph)CH_2COCH_3$$
$$+$$
$$PhCH_2CH(COCH_3)_2$$

(1)

10.4. Carbon–heteroatom bond formation by radical addition

The mechanism of addition of heteroradicals, $R_h\cdot$, to unsaturated compounds is similar to that of carbon-centred radicals. The process usually involves a chain reaction, initiated and terminated in the same way as for alkyl radicals, but with the difference that the addition step is frequently reversible:

$$R_h\cdot + E \underset{k_{-p}}{\overset{k_p}{\rightleftharpoons}} R_hE\cdot$$

$$R_hE\cdot + R_hX \overset{k_a}{\rightarrow} R_hEX + R_h\cdot$$

The carbon–heteroatom bond formed in the addition step is comparatively weak, and there may be appreciable decomposition of the adduct radical even at room temperature. Reversible addition has been definitely established for bromine atoms, for organotin radicals, for some aminyl and phosphinyl radicals, and is particularly important for all thiyl radicals (see p. 103).

Application of the steady-state approximation (see chapter 8, p. 63) to the mechanism of addition, including the reverse decomposition, leads to the following expression for the rate of adduct formation:

$$\frac{d[R_hEX]}{dt} = k_p [E] (V_i/k_c)^{\frac{1}{2}} \{1/(1 + k_{-p}/k_a [R_hX])\}$$

Where V_i is the rate of initiation, and k_c is the rate constant for combination of two heteroradicals. The rate of adduct formation depends not only on the rate of addition (k_p), but also on the concentration of the radical source $[R_hX]$ and on the ratio of the rates of decomposition to transfer (k_{-p}/k_a) of the adduct radicals. The rate constant for decomposition increases strongly with temperature, and hence the rate of adduct formation *decreases* with increasing temperature; therefore good yields of adduct are favoured by low temperatures. High concentrations of both alkene and radical source also increase the rate of adduct formation, and low rates of initiation are beneficial in reducing side reactions.

Halogenation of alkenes by hydrogen bromide was studied in the nineteen thirties by Kharasch and Mayo. In the dark the reaction proceeded by an ionic mechanism in which a proton added to the double bond so as to give the more stable carbonium ion. With propylene, for example, 2-bromopropane was formed.

$$CH_3CH = CH_2 + \overset{\frown}{H-Br} \longrightarrow CH_3\overset{+}{C}HCH_3 + Br^-$$

In the presence of ultraviolet light, or with added peroxides, 1-bromopropane was formed instead. Kharasch and Mayo, and independently Hey and Waters, rationalized these observations in terms of a radical-chain reaction in which the active species was a bromine atom. The difference in orientation in the two cases is a consequence of the different active species in the ionic and radical pathways.

$$Br\cdot + E \rightleftharpoons BrE\cdot \qquad \Delta H = -46 \text{ kJ mol}^{-1}$$

$$BrE\cdot + HBr \rightarrow BrEH + Br\cdot \qquad \Delta H = -21 \text{ kJ mol}^{-1}$$

The understanding of these mechanisms marked a milestone in the development of radical chemistry, and gradually led to the acceptance of an increasing number of radical mechanisms for reactions in solution.

The bromine atom addition step has $\Delta H = -46$ kJ mol^{-1} and the transfer step of the adduct radical with hydrogen bromide has $\Delta H = -21$ kJ mol^{-1}. Since both steps are exothermic the reaction proceeds rapidly. For hydrogen chloride the addition step is exothermic ($\Delta H = -109$ kJ mol^{-1}) but the transfer step is endothermic ($\Delta H = +21$ kJ mol^{-1}), and with hydrogen iodide, although the transfer step is exothermic ($\Delta H = -113$ kJ mol^{-1}) the addition step is endothermic ($\Delta H = +29$ kJ mol^{-1}). Only hydrogen bromide has both steps exothermic and this explains why radical addition is facile for this molecule but unknown for the other hydrogen halides.

The addition of bromine atoms to alkenes is reversible at room temperature and below, and shows an apparently negative temperature coefficient. At temperatures low enough to halt the reverse reaction, hydrogen bromide addition is found to be nearly stereospecific with *cis*-, and *trans*-alkenes, and cycloalkenes (see p. 75).

The molecular halogens, fluorine, chlorine and bromine, also add to alkenes by a radical chain reaction to give dihalides as the main products. The mechanism of initiation is not clear, and it may involve two simultaneous processes. Spontaneous reaction of the molecular halogen with the alkene,

$$X_2 + E \rightarrow XE\cdot + X\cdot$$

and initiation by halogen atoms generated thermally or photochemically.

$$X_2 \xrightarrow[\text{or } \Delta]{h\nu} 2X\cdot$$

$$X\cdot + E \rightarrow XE\cdot$$

In solution the radical reaction competes with ionic processes which are difficult, if not impossible to exclude.

With alkenes containing alkyl substituents, halogen atoms will readily abstract hydrogen from allylic positions. At high temperatures, and with low halogen concentrations this abstraction process predominates and allyl halides can be synthesized in good yield (see chapter 8, p. 76). Even at room temperature and in the presence of high concentrations of halogen, the abstraction reactions may be important. For example, chlorination of but-1-ene gave 85.5 per cent of $CH_3CH_2CHClCH_2Cl$ together with the following abstraction products:

$CH_3CH=CHCH_2Cl$ (75%), $CH_3CHClCH=CH_2$ (2.5%) and

$CH_2ClCH_2CH=CH_2$ (4.5%).

Organosilanes, organogermanes and organotin hydrides add to alkenes and other unsaturated molecules by a radical-chain process. The overall result of the reaction is as follows:

$$R_3MH + XCH=CH_2 \rightarrow R_3MCH_2CH_2X$$

Here R may be an alkyl, aryl, or halogen substituent. Best yields are obtained with terminal alkenes and fluorinated alkenes; 1,2-disubstituted alkenes react slowly, if at all. The usefulness of these reactions in the synthesis of carbon–silicon, carbon–germanium, and carbon–tin bonds is illustrated by the following examples.

$$HSiCl_3 + Bu^tC\equiv CH \xrightarrow{ROOR} \begin{array}{c} Bu^t \quad SiCl_3 \\ \diagdown \quad \diagup \\ C=C \\ \diagup \quad \diagdown \\ H \quad H \end{array} \quad (39\%)$$

$$HGeCl_3 + CH_2=CHCN \rightarrow Cl_3GeCH_2CH_2CN \quad (53\%)$$

The addition of tin hydrides is reversible, but the consequences of this are minimized by the great rapidity of the chain-transfer step, which involves abstraction of hydrogen from the tin hydride by the adduct radical. So rapid is this step that telomers or polymers are not formed in detectable quantities.

$$HSnMe_3 + CH_2=C=CH_2 \rightarrow Me_3SnC(CH_3)=CH_2 + Me_3SnCH_2CH=CH_2$$
$$(30\%) \qquad\qquad\qquad (37\%)$$

A variety of reagents, including dinitrogen tetroxide, nitryl chloride, nitrosyl chloride and tetrafluorohydrazine, will add to alkenes with the formation of new carbon–nitrogen bonds. Synthetically, dinitrogen tetroxide is probably the most important, and this reagent furnishes dinitroalkanes and nitro-nitrites with almost any alkene. The latter products are usually hydrolysed to nitroalcohols in the work-up procedure. The reaction with norbornene is given as an example.

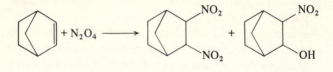

exo-cis 16–20%   *trans* 33–40%

trans   12–14%

Carbon–phosphorus bonds can be formed by the radical addition of compounds of the type R_2PH. Phosphines and derivatives of phosphorous, hypophosphorous and phosphinic acids are all suitable. Examples are as follows:

$Me_2PH + CFCl{=}CF_2 \rightarrow CHFClCF_2PMe_2$ (61%)

$HP(O)(OEt)_2 + MeCO_2CH{=}CH_2 \rightarrow MeCO_2CH_2CH_2P(O)(OEt)_2$ (14%)

$PhPH(O)OEt + C_6H_{13}{}^{n}CH{=}CH_2 \rightarrow C_8H_{17}{}^{n}P(Ph)(O)OEt$ (37%)

Phosphorus trihalides also add to alkenes in a radical process, but oxidized products are obtained unless special precautions are taken to exclude moisture and air.

$$PCl_3 + MeCH{=}CHMe \xrightarrow{\text{air}} MeCHClCHMePOCl_2$$

Relatively few oxygen-containing compounds add to alkenes to give carbon–oxygen bonds. With alcohols and acids the usual methods of initiation lead to the breaking of a C—H bond, because this is weaker than the O—H bond, and a radical centred on carbon is formed. Small yields of adducts can be obtained from the reaction of t-butyl hypochlorite with alkenes, but in general alkoxy radicals add relatively slowly to double bonds, and the main products come from hydrogen abstraction at the side chain of the alkene, e.g.

$$Me_2C{=}CH_2 + Bu^tOCl \rightarrow CH_2ClCMe{=}CH_2 + Bu^tOCH_2CClMe_2$$

$$(83\%) \qquad\qquad (17\%)$$

Perfluoroalkoxy radicals are an exception to this rule and they add readily to a variety of alkenes.

Thiols are very useful radical sources for the formation of carbon–sulphur bonds. The mechanism is similar to that already described, and the addition step is reversible. Preparations are usually carried out with high thiol concentrations and at low temperatures to reduce the extent of reversal and minimize telomerization. Good yields of sulphides can be obtained with alkyl, aryl, and heterocyclic thiols, e.g.

$$EtSH + CH_2 = CHOBu^n \rightarrow EtSCH_2 CH_2 OBu^n$$

Thiol acids also add to a range of unsaturated compounds:

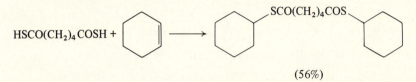

$$HSCO(CH_2)_4 COSH +$$

(56%)

Other sulphur compounds, including sulphonyl and sulphuryl halides, disulphides, bisulphites and sulphur chloride pentafluoride, have been used in essentially the same way.

$$PhSO_2 Br + CH_2 = CHBr \rightarrow PhSO_2 CH_2 CHBr_2 \quad (67\%)$$

$$NaHSO_3 + C_5 H_{11}{}^n CF = CF_2 \rightarrow C_5 H_{11}{}^n CHFCF_2 SO_3 Na \quad (73\%)$$

$$Bu^n SSBu^n + MeC \equiv CH \xrightarrow{h\nu} \underset{\underset{Bu^nS}{}{\overset{Me}{}}}{C} = \underset{\underset{H}{}{\overset{SBu^n}{}}}{C} \quad + \quad \underset{\underset{Bu^nS}{}{\overset{Me}{}}}{C} = \underset{\underset{SBu^n}{}{\overset{H}{}}}{C}$$

(64%)   (16%)

The addition step is less reversible with alkynes, so that the yield of adduct is often greater than with the corresponding alkene.

10.5. Radical fragmentation reactions

The process of radical fragmentation is essentially the reverse of radical addition.

$$R - X - Y \cdot \rightarrow R \cdot + X = Y$$

In certain instances involving heteroradicals (see p. 97) the addition reaction is a reversible process, although equilibrium is rarely obtained owing to the intrusion of other reactions.

$$R_h CH_2 CHX \cdot \rightleftharpoons R_h \cdot + CH_2 = CHX$$

The addition reaction results in a significant decrease in entropy and is thus favoured at low temperatures, whereas the fragmentation process entails an increase in entropy and is thus favoured at high temperatures. In a reversible addition, the amount of reverse decomposition will therefore increase at higher temperatures and the observation of a *decrease* in the rate of adduct formation with increasing temperature is diagnostic of this type of process.

Reversible addition also leads to isomerization of 1,2-disubstituted alkenes. If a radical adds to a *cis*-1,2-disubstituted alkene, the adduct radical can undergo rotation about the carbon–carbon bond, and thus subsequent decomposition can lead to formation of either the *cis*- or the *trans*-alkene.

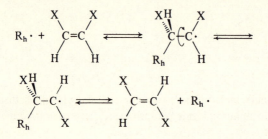

If, therefore, the unconsumed alkene remaining at the end of the reaction is found to be isomerized into a *cis/trans* mixture this is good evidence that the addition is reversible.

β-Haloalkyl radicals can be formed by addition of halogen atoms to alkenes, or by addition of other radicals to haloalkenes. Iodine and bromine are very readily lost from these radicals, and chlorine with somewhat greater difficulty. Bromine and iodine bring about the isomerization of 1,2-disubstituted alkenes in the presence of ultraviolet light. The consequences of this fragmentation are illustrated by the reaction of allyl bromide with excess bromotrichloromethane. None of the expected adduct was detected, and instead, 3-bromo-1,1,1,5,5,5-hexachloropentane (2) was formed together with other products. The formation of this molecule can be explained by the sequence of reactions outlined below:

$$CCl_3 \cdot + CH_2 = CHCH_2 Br \rightarrow Cl_3 CCH_2 \dot{C}HCH_2 Br$$

$$Cl_3 CCH_2 \dot{C}HCH_2 Br \rightarrow Cl_3 CCH_2 CH = CH_2 + Br \cdot$$

$$Cl_3 CCH_2 CH = CH_2 + CCl_3 \cdot \rightarrow Cl_3 CCH_2 \dot{C}HCH_2 CCl_3$$

$$Cl_3 CCH_2 \dot{C}HCH_2 CCl_3 + CCl_3 Br \rightarrow Cl_3 CCH_2 CHBrCH_2 CCl_3 + CCl_3 \cdot$$

(2)

The rapid fragmentation of β-haloalkyl radicals is put to good use in the allylic halogenation of alkenes (see chapter 8, p. 76).

The extent of reversibility in the addition of thiyl radicals to alkenes is greater the more stabilized is the adduct radical. Thus, there is a greater degree of reversibility in the addition of thiols to alkenes than to alkynes, as the alkyl radicals are more stabilized than the vinyl radicals. The β-elimination of thiyl radicals is seen in other reactions which proceed via β-thioalkyl radicals. For example, the peroxide-induced reaction of β-hydroxysulphides results in the formation of carbonyl compounds and thiols.

$$\text{MeCH(OH)CH}_2\text{SMe} + \text{Bu}^t\text{O} \cdot \rightarrow \text{Me}\dot{\text{C}}\text{(OH)CH}_2\text{SMe} + \text{Bu}^t\text{OH}$$

$$\text{Me}\dot{\text{C}}\text{(OH)CH}_2\text{SMe} \rightarrow \text{MeC(OH)}=\text{CH}_2 + \text{MeS} \cdot$$

$$\text{MeC(OH)}=\text{CH}_2 \rightarrow \text{MeCOMe}$$

$$\text{MeS} \cdot + \text{MeCH(OH)CH}_2\text{SMe} \rightarrow \text{Me}\dot{\text{C}}\text{(OH)CH}_2\text{SMe} + \text{MeSH}$$

In reactions involving the addition of thiyl radicals to allyl halides, the resultant radical could undergo two different modes of fragmentation depending on whether a halogen atom or a thiyl radical is split off:

$$\text{RS} \cdot + \text{CH}_2=\text{CHCH}_2\text{X} \longrightarrow \text{RSCH}_2\dot{\text{C}}\text{HCH}_2\text{X} \begin{cases} \text{RSCH}_2\text{CH}=\text{CH}_2 + \text{X} \cdot \\ \\ \text{CH}_2=\text{CHCH}_2\text{X} + \text{RS} \cdot \end{cases}$$

In practice it is found that halogen atoms are generally lost more readily than thiyl radicals.

The reverse of the addition reaction is not observable with alkyl radicals at moderate temperatures. Thus, isomerization of *cis*-2-butene does not occur in its reaction with trichloromethyl or trifluoromethyl radicals. Elimination reactions of alkyl radicals can become important at higher temperatures, and when the process leads to a decrease in strain, as in the ring opening of cyclopropylmethyl radicals (see p. 106 and 163). Similarly, radical polymerization must be carried out below the 'ceiling temperature'. Above this temperature the propagation steps are appreciably reversible and clean polymerization does not occur.

Some of the best known radical fragmentation reactions involve radicals not generally produced by addition reactions. t-Alkoxy radicals are usually made from peroxides or alkyl hypochlorites. They decompose to give alkyl radicals and a ketone.

$$\text{Me}_3\text{CO} \cdot \rightleftharpoons \text{Me}_2\text{CO} + \text{Me} \cdot$$

In the case of differently substituted radicals the process proceeds by breaking the weakest bond, which leads to formation of the most stable radical. Fragmentation leading to loss of hydrogen atoms or aryl radicals does not occur,

$$
\begin{array}{ccc}
\text{Me} & & \text{Me} \\
| & & | \\
\text{Pr}^i\text{-C-OCl} & \xrightarrow{80°\text{C}} & \text{Pr}^i\text{-C-O} \cdot \\
| & & | \\
\text{Et} & & \text{Et}
\end{array}
\xrightarrow{\text{CCl}_4}
\begin{array}{l}
\xrightarrow{(95\%)} \text{MeCOEt} + \text{Pr}^i\text{Cl} \\
\\
\xrightarrow{(3\%)} \text{MeCOPr}^i + \text{EtCl} \\
\\
\xrightarrow{(<0.5\%)} \text{EtCOPr}^i + \text{MeCl}
\end{array}
$$

but intramolecular fragmentation readily takes place when two of the alkyl groups are joined as part of a cycloalkyl system:

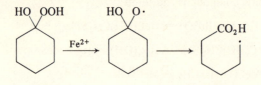

$$2 \cdot CH_2(CH_2)_4CO_2H \longrightarrow HO_2C(CH_2)_{10}CO_2H$$

Acyl radicals are most commonly formed by abstraction of the weakly bound formyl hydrogen from aldehydes, and by photodecomposition of ketones. They readily undergo decarbonylation, and this decomposition competes with their reaction with solvent at moderate temperatures.

$$RCHO \xrightarrow{\cdot CCl_3} R\dot{C}O \underset{CCl_4}{\overset{}{\diagdown}} \begin{array}{l} R\cdot + CO \\ \\ RCOCl + \cdot CCl_3 \end{array}$$

Acyloxy radicals are formed by thermal decomposition of alkanoyl (and aroyl) peroxides and peresters. Loss of carbon dioxide from these radicals is such a rapid process that they have little if any existence outside the solvent cage. The lifetime of the acetoxy radical, for example, has been estimated to be only 10^{-9} to 10^{-10} s.

$$RCO_2 \cdot \rightarrow R \cdot + CO_2$$

Aroyloxy radicals such as $PhCO_2 \cdot$ have much longer lifetimes than acetoxy radicals.

10.6.   Stereochemistry of addition

Radical addition to 1,2-disubstituted alkenes is generally found to be non-stereospecific because the adduct radical is planar and can be approached from either side by the transfer agent. For example, addition of methane thiol-D to *cis*- and *trans*-but-2-ene gave the same mixture of *threo*- and *erythro*-3-deuterio-(2-methylthio)butanes.

The addition of HBr to alkenes is a notable exception to this general rule. Hydrogen bromide (and thiyl radical) addition is reversible at room temperature and therefore the alkene is isomerized in the course of the reaction. However, provided the reaction is carried out at a sufficiently low temperature (*c.* −70°C) and in a great enough excess of HBr to prevent the reverse reaction, almost completely stereospecific product formation can be achieved. Addition of DBr to *cis*-but-2-ene at −70°C gave more than 90 per cent yield of *threo*-3-

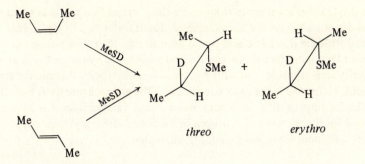

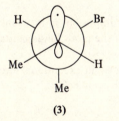

threo   *erythro*

deuterio-2-bromobutane, and a similar yield of the *erythro*-isomer was
obtained from the *trans*-alkene.

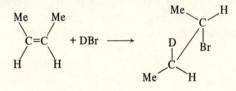

cis-but-2-ene   95% *threo*-3-deuterio-2-bromobutane

The stereospecificity of this reaction can probably be attributed to the preference shown by the intermediate β-bromoalkyl radical for the staggered conformation (3) in which the radical centre is non-planar (see chapter 5, p. 41):

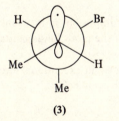

(3)

Radical addition to cycloalkenes is frequently a highly stereoselective
process. Thiyl radicals, silyl radicals, nitroxides and especially bromine atoms,
all show a marked preference for *trans*-addition leading, in the case of 1-
substituted cyclohexenes, to the *cis*-product. For example, in the addition of
thioacetic acid to 4-t-butyl-1-methylcyclohexene, the major product was (4).

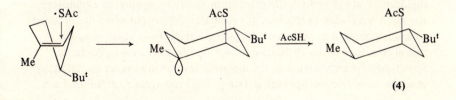

(4)

The thiyl radical approaches from above the plane of the double bond in an axial direction to give the adduct radical. The transfer step of the adduct radical with the thioacetic acid is also axial so that the major product is formed by a *trans*-diaxial process. Radical addition of hydrogen bromide is virtually stereospecific, and with 1-bromo- or 1-chlorocyclohexane the *cis*-1,2-dihalocyclohexane is the exclusive product. The stereochemistry is basically similar for rings of other sizes, and predominant *trans*-addition has been established for the addition of hydrogen bromide to 1-bromocyclobutene, 1-methylcyclopentene and 1-methylcycloheptene.

10.7.   Radical cyclization reactions

Alkenyl radicals of the type $\cdot\,CH_2(CH_2)_n\,CH{=}CH_2$ can cyclize by a process of intramolecular addition. The product may be either a cycloalkylmethyl radical or a cycloalkyl radical, depending on the chain length and the type of substituents present:

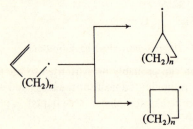

Radical cyclizations are becoming increasingly important in mechanistic and synthetic organic chemistry, and a number of the more fundamental examples will be considered in this section.

Intramolecular addition in homoallyl radicals can, in exceptional circumstances, lead to the formation of cyclopropylmethyl radicals.

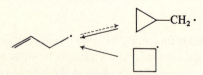

Normally, the reverse reaction, in which the cyclopropylmethyl radicals fragment, is dominant, and the equilibrium lies far to the left. Thus, when homoallyl radicals are generated, the reaction products are mainly, or exclusively, acyclic. A significant proportion of cyclization does occur when the cyclized radical is stabilized by, for example, phenyl substituents adjacent to the radical centre, as in the cyclization of 4,4-diphenylbut-2-en-1-yl radicals. Sometimes the architecture of the molecule favours intramolecular addition, as in the norborn-5-en-3-yl radical (see p. 163) and again cyclized products

may form an appreciable proportion of the total. Cyclization of homoallyl
radicals to cyclobutyl radicals has never been observed, although cyclobutyl radicals fragment into homoallyl radicals.

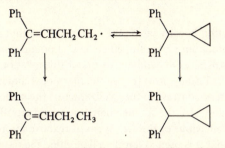

Pent-4-en-1-yl radicals also cyclize rather reluctantly. No cyclized products
are obtained when these radicals are generated under conditions in which
hex-5-en-1-yl radicals cyclize more or less quantitatively. Significant yields of
cyclized products are only obtained when the radical contains structural or
electronic features which favour this type of behaviour, or when the acyclic
radical is stabilized by suitable substituents which extend its lifetime. When
cyclization occurs it invariably gives cyclopentyl radicals, and never the cyclo-
butylmethyl radicals.

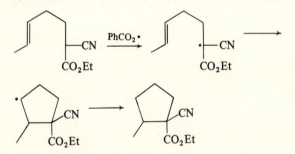

Hex-5-en-1-yl radicals generated in a variety of ways undergo rapid intramolecular addition to give cyclized products:

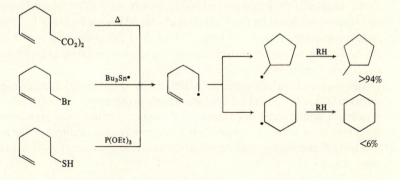

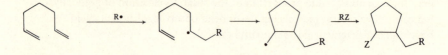

The main product of the reaction in each case is the cyclopentylmethyl radical, but this is accompanied by minor amounts of open-chain compounds and traces of the cyclohexyl radical. The outstanding feature of this reaction is that the parent hex-5-en-1-yl radical cyclizes to give the thermodynamically less stable radical. This is in stark contrast to the cyclization of the analogous carbonium ions, which invariably give the six-membered ring products. The difference in the cyclization behaviour of the two types of reactive inter-mediates is so characteristic that it is used as a test to determine the nature of the reactive species in mechanistic investigations.

The preference shown by hex-5-en-1-yl radicals for formation of five-membered rings may be a result of the geometry of the transition state. The transition state for radical cyclization probably involves interaction of the unpaired electron with the lowest unfilled orbital of the π-system. Conse-quently, the approach of the radical centre occurs preferentially from above (or below) the plane of the unsaturated section and along a line extending almost vertically from one of the carbon atoms of the double bond (5).

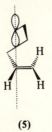

(5)

It follows from this that bond formation will occur at the terminus of the alkenic linkage most readily approached by the radical centre from vertically above. Models of the hex-5-en-1-yl radical clearly show that five-membered ring formation should be favoured, though the difference between the five- and six-membered rings is not large. This explanation is also in accord with the cyclization of hept-6-en-1-yl radicals to cyclohexylmethyl and with the reluctance of pent-4-en-1-yl radicals to cyclize.

Hex-5-en-1-yl radicals which are stabilized by electron-withdrawing substi-tuents attached to the radical centre are more prone to give cyclohexane derivatives, and in some cases these are the major products. The cyclization of these radicals is reversible which leads to thermodynamic control of the pro-cess and hence the thermodynamically more stable cyclohexyl radical is the main product.

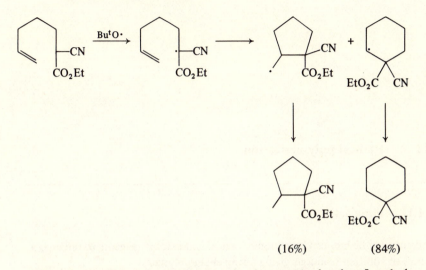

(16%)   (84%)

Alkenyl radicals with longer chains cyclize less readily than hex-5-en-1-yl radicals, and a greater proportion of open-chain products is obtained in their reactions. The cyclization process leads to the preferential formation of the cycloalkylcarbinyl radical. Thus, hept-6-en-1-yl and oct-7-en-1-yl radicals generated from the alkenyl bromides by treatment with tri-n-butyltin hydride cyclize to give cyclohexane and cycloheptane derivatives respectively.

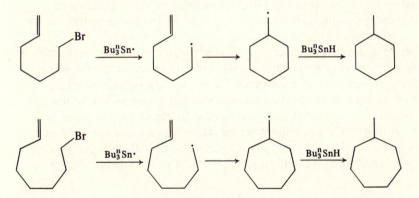

The yields of cyclized products can be increased by using acyclic radicals stabilized with electron-withdrawing substituents adjacent to the unpaired electron.

Extensive studies have been made on aromatic systems in which cyclization occurs by intramolecular attack of alkyl or aryl radicals on an aromatic ring. This kind of intramolecular homolytic aromatic substitution occurs with 4-phenylbutyl and related radicals, and in the Pschorr reaction (see chapter 12, p. 129).

11 Radical polymerization

11.1. Introduction

Alkenes can be induced by cationic, radical and anionic reagents to undergo a series of addition reactions giving a long-chain polymer.

$$n\mathrm{CH_2{=}CHX} \rightarrow -\mathrm{CH_2(CHX{-}CH_2)}_{n-1}\mathrm{CHX}-$$

The active end of the growing polymer may thus be a carbonium ion, a radical or a carbanion, and the polymer grows by a chain reaction. This type of polymerization is called 'addition polymerization' or alternatively 'vinyl polymerization'. Electron-withdrawing substituents in the monomer favour propagation by anionic species and electron-releasing substituents encourage initiation by cations. Radicals, being uncharged are less discriminating and are thus useful initiators for most kinds of monomers.

In a polymerization initiated by radicals, the process involves rapid addition of monomer to a few active centres. Even at low conversion the mixture consists of high polymer and unreacted monomer. The polymer produced contains a linear sequence of repeat units identical in composition to the monomer. The polymer chain is held together by carbon–carbon bonds, and the substituents (X) in the monomer are attached to this chain as 'pendant' groups. If the substituent contains an unsaturated bond then further polymer growth can occur at the pendant group, thus leading to chain branching.

11.2. Mechanism of radical polymerization

Radical polymerization shows all the characteristics of a chain reaction. It is initiated by peroxides, azo-compounds, electron-transfer reagents or ultra-violet light, and it is susceptible to inhibition by radical inhibitors. In the photochemical process, the quantum yield, expressed as the number of molecules polymerized per quantum absorbed, can be as high as 10^3 and molecular weights of the resulting polymers can be 10^5 and higher.

Decomposition of the initiator (In) produces radicals which add to the monomer (M) to give an active centre. Many initiators including peroxides and azo-alkanes, decompose to give two radicals; consequently, two polymer chains are started for each initiator molecule decomposed.

$$
\left.
\begin{aligned}
\text{In} &\xrightarrow{k_i} 2R\cdot \\
R\cdot + M &\to RM_1\cdot
\end{aligned}
\right\}
\quad \textit{initiation}
$$

The growth of the polymer radical occurs by a series of rapid additions of monomer units:

$$
\left.
\begin{aligned}
RM_1\cdot + M &\xrightarrow{k_p} RM_2\cdot \\
&\;\cdot \\
&\;\cdot \\
&\;\cdot \\
RM_r\cdot + M &\xrightarrow{k_p} RM_{r+1}\cdot
\end{aligned}
\right\}
\quad \textit{propagation}
$$

Chain termination usually occurs by combination or disproportionation of polymer radicals:

$$
\left.
\begin{aligned}
RM_r\cdot + RM_r\cdot &\xrightarrow{k_c} RM_{r+r}R \\
RM_r\cdot + RM_r\cdot &\xrightarrow{k_d} RM_r H + RM_r(-H)
\end{aligned}
\right\}
\quad \textit{termination}
$$

The propagation and termination rate constants k_p and k_t $(= k_c + k_d)$ will in general depend on the length of the polymer radical involved. However, there is good experimental evidence that both these rate constants are independent of chain length for polymer radicals incorporating more than a few monomer units. The various polymer radicals $RM_r\cdot$ are therefore kinetically indistinguishable, and we can use k_p and k_t as general rate constants for all the propagation or termination steps.

The rate of consumption of monomer, which is the rate of polymerization V_p is given by:

$$
-d[M]/dt = V_p = k_p[RM_r\cdot][M]
$$

During the steady-state propagation of the polymerization the rate of initiation (V_i) is equal to the rate of termination:

$$
V_i = 2k_t[RM_r\cdot]^2
$$

(the factor 2 arises because identical radicals are involved in termination). Substituting this into the equation for the rate of polymerization gives:

$$
V_p = k_p[M](V_i/2k_t)^{\frac{1}{2}}
$$

When the polymerization is initiated by thermal decomposition of a peroxide or azo-compound the rate of initiation is given by $V_i = 2k_i f[In]$, where k_i is the rate constant of initiator decomposition, f is the fraction of initiator radicals which actually start chains and the factor 2 allows for the fact that two radicals are formed from each initiator molecule. In photochemically initiated polymerizations $V_i = \phi I_a$, where ϕ is the quantum yield of chain initiation and I_a is the absorbed light intensity. The rate of polymerization therefore becomes:

$$V_p = k_p (k_i f [\text{In}]/k_t)^{\frac{1}{2}} [\text{M}] \quad \text{or} \quad V_p = k_p (\phi I_a/k_t)^{\frac{1}{2}} [\text{M}]$$

This simple kinetic analysis is based on the assumptions that: termination occurs solely by mutual interaction of two polymer radicals; that no chain transfer takes place; that the polymer radicals have the same reactivity irrespective of their structure; and that chain branching, inhibition and other complications are absent.

Many radical polymerizations do obey this simple relationship. In the polymerizations of methyl methacrylate and styrene initiated by azo-bisiso-butyronitrile (AIBN) and in the dibenzoyl peroxide-initiated polymerization of styrene, methyl methacrylate, vinyl acetate, and (+)-*sec*-butyl-α-chloro-acrylate, it has been confirmed that the rate of polymerization is directly proportional to the square root of the initiator concentration. The rate of polymerization is also found to be proportional to the monomer concentration, except for those systems in which the efficiency of initiation, f, varies with monomer concentration. This occurs when induced decomposition of the initiator sets in, or when cage combination of the initiator radicals is important. In the polymerization of methyl methacrylate using a series of different azoalkanes as initiators, the rates of polymerization were shown to be proportional to the square roots of the rate constants of initiator decomposition (k_i).

The kinetic chain length, ν, which represents the number of monomer units reacting with each polymer radical during its lifetime, is given by the ratio of the rate of propagation to the rate of termination, i.e.:

$$\nu = V_p/V_i = (k_p/2k_t)[\text{M}]/[\text{RM}_r \cdot]$$

On substitution for the polymer radical concentration from the equation for the rate of monomer consumption this becomes:

$$\nu = (k_p^2/2k_t)[\text{M}]^2/V_p$$

The chain length is inversely proportional to the polymer radical concentration, hence the chain is short for high radical concentrations and vice versa. The polymer molecular weight can thus be controlled to some extent by varying the initiator concentration. An increase in the temperature increases the rate of polymerization, V_p, but decreases the chain length.

11.3. Initiation

Photochemical initiation offers some advantages for kinetic studies, and for other applications where rapid and close control of the energy input is required. For large-scale preparative purposes thermal initiation with peroxides or azo-compounds is generally preferred. Redox systems such as iron(II) sulphate plus hydrogen peroxide or peroxydisulphate are also valuable.

$$Fe^{2+} + H_2O_2 \rightarrow Fe^{3+} + OH^- + \cdot OH$$

$$Fe^{2+} + S_2O_8{}^{2-} \rightarrow Fe^{3+} + SO_4{}^{2-} + SO_4{}^{-}$$

Transition-metal carbonyls such as $Mo(CO)_6$ and related organometallic complexes are also coming into use.

The primary radicals produced in the decomposition of the initiator may not all start polymer chains. The fraction of initiator radicals which actually start chains, f, gives a measure of the efficiency of initiation. The efficiency factor, f, can approach unity for some systems, but it is usually reduced by side reactions. The primary radicals may undergo cage combination before they can diffuse into the solution and react with monomer e.g.:

$$Me_2(CN)C-N=N-C(CN)Me_2 \xrightarrow{\Delta} [Me_2(CN)C \cdot \quad N_2 \quad \cdot C(CN)Me_2]_{cage}$$

combination / \ *diffusion*

$$Me_2(CN)CC(CN)Me_2 \qquad 2Me_2(CN)C \cdot$$

The extent of cage combination depends on the viscosity of the solution, the temperature, and other factors. The number of primary radicals available for chain initiation is also reduced when the initiator undergoes induced decomposition by radical attack (see p. 188):

$$R' \cdot + ROOR \rightarrow R'OR + RO \cdot$$

Effectively, this reaction reduces the number of primary radicals produced per molecule of initiator from two to one. Because of these two reactions the efficiency factor depends on the conditions of polymerization as well as on the initiator itself. For most initiation systems f lies between 0.8 and 0.3.

11.4. Propagation

The chain-propagation step involves addition of polymer radicals to the monomer. The reactivity of polymer radicals varies very widely. For mono- or 1,1-disubstituted alkenes and for dienes the polymerization is a very exothermic process and the propagation step is therefore very fast and not reversible at normal temperatures. Polymerizations involving aromatic monomers, and those in which the chain-carrying radicals have the unpaired electron centred on a heteroatom (as in polymerizations with sulphur dioxide), can be reversible even at room temperature.

The structures of both the monomer and the polymer radical influence the rate of the propagation reaction. The most important factors which control the rate of the reaction appear to be (i) polar, and (ii) steric (see chapter 10, p. 90). It is found that monomers can be arranged in order of reactivity corresponding to the polarity of their substituents. The greater the electronegativity of the substituents the slower is the rate of polymerization. The reactivity of the monomers correlates with the Hammett σ-constants of their substituents. That steric effects are important is shown by the reluctance of most 1,2-disubstituted alkenes to polymerize. When some tendency to polymerize is found for a 1,2-disubstituted alkene the *cis*- and *trans*-isomers are usually found to have quite different reactivities. This behaviour can also be attributed to steric effects of the substituents. Steric effects are probably negligible for monomers which have one end of the double bond unsubstituted.

The unsymmetric nature of most monomers means that there are two ways in which the growing polymer radical can react with them:

$$RM_r \cdot + CH_2 {=} CHX \rightarrow RM_r CH_2 CHX \cdot$$

$$RM_r \cdot + XCH {=} CH_2 \rightarrow RM_r CHXCH_2 \cdot$$

Radical addition nearly always occurs preferentially at the less substituted end of an alkene because steric hindrance is smaller (see chapter 10). The first of these reactions will in general be favoured so that the final polymer should consist of a regular head-to-tail sequence of monomer units. This head-to-tail arrangement of monomers has been found for all vinyl polymers so far studied.

In vinyl polymers of the type $RM_r CH_2 CXYM_s R$, the carbon atom of the CXY unit will be chiral, and two stereoisomeric configurations will be possible. These (R) or (S) configurations may be distributed in various ways along the polymer chain. Polymers with a high degree of stereoregularity are said to be *tactic* and those with a random distribution are termed *atactic*. The radical polymerization process virtually always produces polymer which is neither tactic nor atactic, but consists of short tactic sequences separated by atactic units.

11.5. Termination

The most common form of chain termination is combination or disproportionation of two polymer radicals. The example of methyl methacrylate is shown at the top of the next page.

At present there is no simple way of predicting from the structure of the polymer radicals whether combination or disproportionation will be favoured. It appears that higher temperatures favour disproportionation. For simple alkyl radicals the disproportionation : combination rate constant ratios are high for branched tertiary and secondary radicals (see chapter 7). It seems

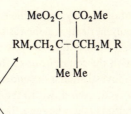

$$2RM_rCH_2\overset{\cdot}{C}MeCO_2Me$$

$$RM_rCH_2C-CCH_2M_rR$$

(top) MeO$_2$C   CO$_2$Me
| |
RM$_r$CH$_2$C−CCH$_2$M$_r$R
| |
Me Me

(bottom) RM$_r$CH$_2$CHMeCO$_2$Me + RM$_r$CH=CMeCO$_2$Me
+ RM$_r$CH$_2$CCO$_2$Me
‖
CH$_2$

probable that this rule may also hold for polymer radicals, and the observation that disproportionation is favoured by methyl methacrylate, but that styrene and acrylonitrile terminate mainly by combination, lends support to this view.

Termination processes which are first order in polymer radical concentration have also been observed. It is probable that this kind of termination occurs when the polymer radical attacks the monomer to give a radical which is too unreactive to initiate further chains. This occurs for instance with allyl monomers where the polymer radical can abstract an allylic hydrogen giving a resonance-stabilized radical which does not add to monomer.

$$RM_r\cdot + CH_2=CHCH_2X \longrightarrow RM_rH + \overset{\cdots\cdots\cdots}{CH_2-CH-CHX}$$

The rate of termination by combination or disproportionation depends on the square of the polymer radical concentration, and therefore long chains leading to high molecular weight polymer are favoured by low radical concentrations. The termination rate constant can be measured by non-stationary state techniques such as the intermittent illumination method. For most polymer radicals k_t is found to be in the range 10^6 to 10^8 l mol^{-1} s^{-1}. For two polymer radicals to interact with each other they must first diffuse together in the solution so that the active ends come into close proximity. This process in which the centres of gravity of the polymer chains approach each other is called translational diffusion. If the polymer chains are long, the radical sites may still be separated from each other by several solvent molecules or inert segments of the chains. The radical centres may then approach close enough for chemical reaction as each chain moves through a variety of conformations and this is called segmental rotation.

It has recently become apparent that the termination reaction of most polymerizations is controlled by the physical rate at which the radical centres diffuse together and that the chemical process of combination or disproportionation is comparatively rapid. The rate constant of a diffusion-controlled reaction should depend on the inverse of the solution viscosity (see chapter 7) and since the rate of polymerization depends on $(k_2)^{-1/2}$, i.e. $V_p = k_p[M]$ $(V_i/2k_t)^{1/2}$ (see above), it follows that the rate of polymerization would be expected to increase with the square root of the viscosity. In all systems which have so far been investigated a dependence of polymerization rate on viscosity has been observed, even down to the lowest attainable viscosities.

In many polymerizations taken beyond the first few per cent reaction a marked increase in the rate of reaction is observed, instead of the decrease in reaction which would be expected from the depletion of monomer and initiator. This auto-acceleration is known as the *Trommsdorff-Norrish* effect or *gel effect*, and it is a consequence of the increased viscosity of the medium. As the amount of polymer builds up in the solution the active ends of the polymer radicals are prevented from diffusing together by extensive entanglements. The termination rate is dramatically reduced by one or even two orders of magnitude. The propagation step is affected far less by the increased viscosity because diffusion of the small monomer units to the radical centres is hardly impeded. Consequently acceleration in the rate is observed until the reactants are entirely consumed.

11.6.   Chain transfer

An important side reaction in many polymerizations is abstraction of hydrogen from reactants and products by the polymer radicals. Abstraction from initiators is not usually important if the initiator concentration is low, but abstraction from the monomer may be important if it contains allylic, benzylic or tertiary hydrogens.

$$RM_r\cdot + CH_3CH=CH_2 \longrightarrow RM_rH + \overline{CH_2-CH-CH_2}$$

If the resulting radical is sufficiently reactive it will initiate new chains, so that the active centre is not destroyed but transferred to the new radical.

Abstraction often occurs at sites on the chain of product polymer molecules. In this case a new radical site is opened up which can then add to further monomer units thus giving a branched polymer. The branching in polyethylene is believed to arise from this mode of transfer. Polyvinyl acetate is highly branched even at low temperatures because the polyvinyl acetate radicals tend to abstract hydrogen from the acetate function in the polymer molecule.

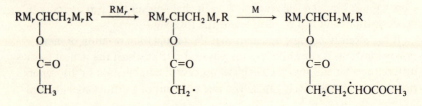

The polymer chain length is reduced by the transfer reaction, and thus the molecular weight of the polymer can be controlled by adding the appropriate amount of a transfer agent RH.

$$RM_r \cdot + RH \xrightarrow{k_a} RM_r H + R \cdot$$

The effectiveness of the transfer agent is measured in terms of the chain-transfer constant C_s which is defined as the ratio of the rate constant for transfer to the rate constant for propagation i.e. $C_s = k_a/k_p$. The rate of transfer is given by:

$$-d[RH]/dt = k_a[RH][RM_r \cdot]$$

and when this is combined with the rate of chain propagation the expression becomes:

$$d[RH]/d[M] = C_s[RH]/[M]$$

which on integration gives:

$$\ln[RH]/\ln[M] = C_s$$

so that the chain-transfer constant can be determined from measurements on the concentration of the monomer and transfer agent during the polymerization. In industrial polymer productions a transfer agent having C_s in the region of unity is usually added to the polymer feed in sufficient quantity to give the desired molecular weight of polymer. It is then consumed at approximately the same rate as the monomer and the ratio $[RH]/[M]$ remains constant throughout the reaction. Aliphatic thiols are frequently suitable for this purpose.

If the added transfer agent or solvent has a high transfer constant, chain transfer may occur so rapidly that the product polymer incorporates only a few monomer units. This type of process is termed telomerization and the small polymers produced are called telomers. In the limiting case when the transfer agent is in excess only one monomer unit may be involved so that the product is the 1 : 1-adduct of the alkene and the transfer agent. Radical additions of this kind are discussed in chapter 10, p. 89.

11.7. Copolymerization

When two or more monomers are mixed together and polymerized the product macromolecule contains units from all the monomers and is called a copolymer. The physical properties of a copolymer of two monomers are usually a compromise between the properties of the homopolymers, and it may retain desirable features of both. The range of practical application of a polymer may therefore be extended by incorporating into it units from more than one monomer. Cationic or anionic copolymerization generally fails to occur or leads to block copolymers consisting of long sequences of the same monomer. This contrasts with the ready copolymerization of a wide variety of monomers by the radical process, although alkenes show surprising differences in their copolymerization behaviour. 1,2-Disubstituted alkenes which do not readily polymerize alone, are capable of facile radical copolymerization with other monomers. Maleic anhydride, for example, readily polymerizes with styrene, and with stilbene, although neither maleic anhydride nor stilbene can give high polymer on its own.

We assume that all polymer radicals having the same radical site are kinetically indistinguishable, i.e. that their reactivities are unaffected by the penultimate monomer unit. There will then be only four important propagation steps in the copolymerization of two monomers M and N:

$$RM_r \cdot + M \xrightarrow{k_{mm}} RM_{r+1} \cdot$$

$$RM_r \cdot + N \xrightarrow{k_{mn}} RM_r N \cdot$$

$$RN_r \cdot + M \xrightarrow{k_{nm}} RN_r M \cdot$$

$$RN_r \cdot + N \xrightarrow{k_{nn}} RN_{r+1} \cdot$$

Application of the steady-state approximation to this system, neglecting the small proportion of polymer radicals involved in initiation and termination reactions and utilizing the assumption that the reactivities of $RM_r \cdot$ and $RN_r M \cdot$ are the same, leads to the following relation:

$$k_{nm} [RN_r \cdot] [M] = k_{mn} [RM_r \cdot] [N]$$

The rates of consumption of the monomers are given by:

$$-d[M]/dt = k_{mm} [M] [RM_r \cdot] + k_{nm} [M] [RN_r \cdot]$$

$$-d[N]/dt = k_{nn} [N] [RN_r \cdot] + k_{mn} [N] [RM_r \cdot]$$

Substitution for $[RN_r \cdot]$ into these equations and division leads to:

$$d[M]/d[N] = \frac{r_1([M]/[N]) + 1}{r_2([N]/[M]) + 1}$$

where $r_1 = k_{mm}/k_{mn}$ and $r_2 = k_{nn}/k_{nm}$. The ratio of the rates of consumption of the two monomers $d[M]/d[N]$ gives the proportions of the two monomers in the copolymer.

The copolymer composition equation as this expression is called, relates the polymer composition to the reactivity ratios r_1 and r_2 of the monomers and the concentration ratio of the monomers in the reaction mixture. If monomer M is more reactive than monomer N then it will be used up more rapidly, and the relative concentrations of the monomers will change; this in turn will affect the copolymer composition which will change throughout the reaction. The monomer reactivity ratios, r, express the preference a polymer radical has to attack its own type of monomer. If $r_1 > 1$ the radical $RM_r \cdot$ prefers to add to M, but if $r_1 < 1$ then $RM_r \cdot$ adds preferentially to N. Experimentally, r values are determined from analysis of the copolymer composition for a series of monomer concentration ratios. The monomer reactivity ratios have been determined for a very large number of monomer pairs, and a selection of the available values are shown in table 11.1.

Table 11.1. *Monomer reactivity ratios from copolymerization studies at 60°C*

M	r_1	N	r_2
Styrene	0.54	Methyl methacrylate	0.42
Styrene	0.30	Methylacrylonitrile	0.16
Styrene	0.40	Acrylonitrile	0.04
Styrene	0.75	Methyl acrylate	0.18
Styrene	55	Vinyl acetate	0.01
Methyl methacrylate	0.67	Methacrylonitrile	0.65
Methyl methacrylate	1.20	Acrylonitrile	0.16
Methyl methacrylate	0.30	Methyl acrylate	1.50
Methyl methacrylate	20	Vinyl acetate	0.02
Methacrylonitrile	2.68	Acrylonitrile	0.32
Methacrylonitrile	12	Vinyl acetate	0.01
Acrylonitrile	1.4	Methyl acrylate	0.8
Acrylonitrile	4.05	Vinyl acetate	0.06
Methyl acrylate	9.0	Vinyl acetate	0.1

The reactivity ratios of a given monomer pair immediately give some information about the structure of the copolymer. Several limiting cases are worth considering: (i) When $r_1 = r_2 = 1$ the two polymer radicals ($RM_r \cdot$ and $RN_r \cdot$) add to either monomer (M or N) with equal facility, and the two types of monomer will be incorporated randomly along the chain. The copolymers formed under these conditions are called *ideal* copolymers. This type of behaviour will occur when the radical site is the same, i.e. carries the same or very similar substituents in the two types of polymer radicals, and when attack occurs at similarly substituted ends of the two alkenes. Similar random incorporation will occur for any monomer pair having $r_1 r_2 = 1$. Approximately ideal behaviour of this kind is shown by the pairs: methyl acrylate + vinyl acetate, and tetrafluoroethylene + chlorotrifluoroethylene. (ii) When $r_1 = r_2 = 0$ the polymer radicals ($RM_r \cdot$) add exclusively to the other type of monomer, (N) and the copolymer will consist of an *alternating* sequence of the two monomers. Alternating copolymers can be prepared from monomer pairs: styrene + maleic anhydride and fumaronitrile + α-methylstyrene. Strictly alternating copolymers are rather rare, but the majority of monomer pairs encountered in practice have $0 < r_1 r_2 < 1$ (see table 11.1) and thus have some tendency to alternate. (iii) When r_1 and r_2 are greater than unity, i.e. $r_1 r_2 > 1$, polymer radicals $(RM_r \cdot)$ add preferentially to their own type of monomer (M) and the copolymer will consist of long sequences or blocks of each monomer. Block copolymer formation of this type is very rare in radical-initiated polymerization; the pair styrene + isobutene is one of the few systems showing this tendency.

The important factors which relate the structures of the reacting species to the rate of the propagation reaction have been referred to above (p. 113) (see also chapter 10, p. 90). The *Q–e scheme* is an attempt to interpret the reactivities of monomers in terms of empirical parameters which characterize the inherent reactivities and the polarities of the monomer M and its associated polymer radical $RM_r \cdot$. The basic equation of the *Q–e* scheme for a cross-propagation reaction is:

$$k_{mn} = P_m Q_n \exp(-e_m e_n)$$

where P_m and P_n characterize the inherent reactivities of the polymer radicals $RM_r \cdot$ and $RN_r \cdot$, Q_m and Q_n relate to the reactivities of the monomers M and N, and e_m and e_n measure the polarities of the monomers M and N which are assumed to be the same as those of the corresponding polymer radicals.

From the basic equation it is easy to show that the monomer reactivity ratios are given by:

$$r_1 = (Q_m/Q_n) \exp[-e_m(e_m - e_n)]$$
$$r_2 = (Q_n/Q_m) \exp[-e_n(e_n - e_m)]$$

The reactivity ratios for a monomer pair can be calculated from the Q and e parameters of the monomers. For a set of i monomers there are approximately $i^2/2$ possible monomer pairs, so that the $Q–e$ scheme offers a valuable short cut around most of the experimental work which would be involved in determining all the reactivity ratios. Styrene has been adopted as the standard monomer and arbitrarily assigned the values $Q = 1.0$ and $e = -0.8$. The Q and e values of all other monomers have been derived from experimental copolymerizations with styrene.

Monomer reactivity ratios calculated from $Q–e$ values are in good agreement with experiment except where steric effects are important, as in copolymerizations involving 1,2-disubstituted alkenes. The $Q–e$ scheme has proved useful for systematizing polymerization data and for estimating the reactivity ratios of monomer pairs for which experimental data have not been obtained. The main drawback to the $Q–e$ scheme is the rather nebulous way in which the Q or e parameters are related to theoretical concepts and to the structures of the monomer and polymer radical.

Numerous attempts have been made to improve the predictive power of the *Q–e* scheme and give it a more rigorous theoretical framework. Probably the most useful extention of the method is the '*patterns of reactivity*' approach in which the polarity of the polymer radical is directly represented by the Hammett σ-constant of the substituent(s) at the radical centre. The rate constant of a propagation or transfer reaction is given by:

$$\log k_s = \log k_{iT} + \alpha\sigma + \beta$$

Where the polymer radical is characterized by σ and k_{iT}, its rate constant for hydrogen abstraction from toluene, while α and β stand for the polar character and the inherent reactivity of the substrate. The monomer reactivity ratios are functions of the σ-constants of the radicals and the α and β values of the monomers i.e.:

$$\log r_1 = \sigma_m(\alpha_m - \alpha_n) + (\beta_m - \beta_n)$$

$$\log r_2 = -\sigma_n(\alpha_m - \alpha_n) - (\beta_m - \beta_n)$$

These equations are formally very similar to those from the *Q–e* scheme. The patterns treatment also gives good reproduction of experimental reactivity ratios. Its main advantages are that it makes separate allowance for the polar character of both the monomer and polymer radical, and that it relates the reactivity ratios to the Hammett σ-constants; thereby interrelating polymer reactivities with a large body of kinetic information from diverse areas of chemistry.

11.8. Principles of polymer synthesis

Radical-initiated polymerizations are much the most important for the gross industrial production of high polymers. Low-density polyethylene, poly(vinyl chloride) and polystyrene are the top three in terms of world annual production, and all three are made by radical processes. Styrene–butadiene copolymer which is currently the major general purpose synthetic rubber, poly(tetrafluoroethylene) and many other vinyl polymers with specialized uses are made by the radical method. High-density polyethylene and polypropylene, which rank fourth and fifth in the world production league are made by non-radical processes using Ziegler–Natta catalysts.

A very wide variety of mono- and 1,1-disubstituted alkenes can be polymerized by radical initiators to give high polymers. Exceptions to this generalization have been mentioned above. 1,2-Disubstituted ethylenes seldom give high molecular weight material in homopolymerization, except when the substituents comprise part of a strained ring system or are alkoxy groups. Monosubstituted alkynes, $RC\equiv CH$, are also more difficult to polymerize and usually give low molecular weight product. Conjugated dienes form polymers with great ease except when they contain substituents in the 1- and 4-positions.

There are four main methods for the radical polymerization of alkenes.
(i) *Bulk polymerization* involves reaction of the initiator with pure monomer. It is usually carried out in two stages, the first to low conversion at low temperature and the second at higher temperature to increase the fluidity of the mixture. The increased viscosity of the solution towards the end of the polymerization promotes the Trommsdorff–Norrish effect, thus ensuring high conversion and high molecular weight polymer. The method presents formidable engineering problems because of the difficulty in handling viscous melts and because the heat of reaction is considerable and must be dissipated to prevent the development of hot spots. The final polymer is of high optical clarity and superior purity and the method is used in some industrial plants for making poly(vinyl chloride), polystyrene, and poly(methyl methacrylate).
(ii) *Solution polymerization*, in which monomer and solvent are mixed together and flow into the reactor, overcomes the problems of heat dissipation and handling of viscous melts, which are associated with the bulk method. The solvent must be chosen with care to prevent the occurrence of chain transfer to solvent. The method is used for ethylene, vinyl acetate and acrylonitrile.
(iii) *Suspension polymerization* procedures employ a water phase which contains suspended droplets of the monomer. The size of the droplets is an important factor in determining the physical properties of the polymer, and it is controlled by adding suspending agents such as poly(vinyl alcohol), gelatine or water-soluble cellulose derivatives. The initiator must be soluble in the droplets and dodecyl peroxide is widely used. This is the most common method for polystyrene production and it is also used for poly(vinyl chloride) and poly(methyl methacrylate).
(iv) *Emulsion polymerization* is extensively used in Europe for production of acrylics, poly(vinyl chloride), poly(vinyl acetate) and styrene–butadiene rubber. As in suspension polymerization, water is the continuous phase which acts as the heat-dispersion medium. Small droplets, or micelles, in which the monomer dissolves are formed by means of the emulsifying agent; which is usually a sulphonate ester of a long chain fatty acid. Water-soluble redox initiators such as persulphate are used. The polymerization takes place in the interior of the micelles which gradually swell throughout the reaction as more monomer and initiator radicals diffuse inside from the aqueous phase.

Commercial polymers contain compounding additives to improve or modify their physical properties. *Stabilizers*, such as lead salts or organotin derivatives confer better thermal stability on the polymer. *Plasticizers* including di-iso-octylphthalate and trixylyl phosphate, increase the flexibility and improve the toughness of the material. *Extenders* are cheap fillers used to replace up to one-third of the more expensive polymer without significantly affecting its properties.

12 Homolytic aromatic substitution

12.1. Introduction

This chapter considers substitution reactions of aromatic compounds in which the attacking species is a radical.

$$R \cdot + ArX \rightarrow ArR + [X \cdot]$$

In the vast majority of cases the group displaced is hydrogen. Homolytic aromatic substitutions are distinguished from electrophilic aromatic substitutions by their relative insensitivity to polar influences either in the substrate or in the attacking radical.

This class of reaction has been very extensively studied, in particular by Hey and his co-workers. Arylation will be considered in most detail because of the large amount of quantitative data pertaining to this reaction and because of its synthetic importance in the preparation of biaryls.

12.2. Mechanism of homolytic aromatic substitution

The initial step in homolytic aromatic substitution involves attack by the radical on the substrate to give an adduct radical. This process is certainly reversible in some instances, as in the reaction of benzoyloxy radicals with benzene, though the question of the reversibility of this step is very much an open question in the attack of phenyl radicals on benzene to give phenylcyclohexadienyl radicals.

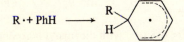

$$R \cdot + PhH \longrightarrow \qquad \qquad (1)$$

The phenylcyclohexadienyl radicals react further in three ways:

(i)   they can disproportionate to give biphenyl and a dihydrobiphenyl (reaction 2).

(ii)   they can dimerize to give a tetrahydroquaterphenyl (reaction *3*).
or
(iii)   they can be oxidized to biphenyl (reactions *4* and *5*).

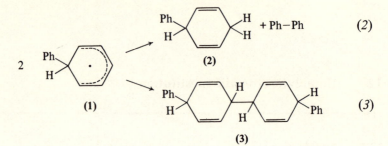

$$[Ph.PhH]\cdot + R\cdot \rightarrow Ph-Ph + RH \tag{4}$$

$$[Ph.PhH]\cdot \xrightarrow{-e^-} [Ph.PhH]^+ \rightarrow Ph-Ph + H^+ \tag{5}$$

Evidence for the intermediacy of phenylcyclohexadienyl radicals comes from the isolation of both 1,4-dihydrobiphenyl (2) and 1′, 1″, 4′, 4″-tetrahydroquaterphenyl (3) from the decomposition of a very dilute solution of dibenzoyl peroxide in benzene: equimolar amounts of biphenyl and dihydrobiphenyl were obtained. In phenylations with dibenzoyl peroxide, decomposition of which gives initially benzoyloxy radicals which subsequently decompose to phenyl radicals, kinetic studies indicate that reaction *4* (where R = Ph or PhCO$_2$) does not occur to any significant extent. This is not unexpected since the steady state concentrations of phenyl and benzoyloxy radicals would be very small. This type of reaction does, however, occur in phenylations using phenylazotriphenylmethane. This latter decomposes to give long-lived triphenylmethyl radicals as well as phenyl radicals: the triphenylmethyl radicals react with phenylcyclohexadienyl radicals to give biphenyl and also dihydro-adducts (reactions *6* and *7*).

$$Ph-Ph \qquad + \qquad Ph_3CH \tag{6}$$

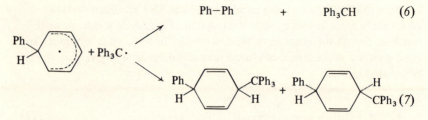

Biphenyl is generated in a chain-transfer reaction with dibenzoyl peroxide.

$$[Ph.PhH]\cdot + (PhCO_2)_2 \rightarrow Ph_2 + PhCOOH + PhCOO\cdot$$

The Gomberg reaction, which involves addition of sodium hydroxide to a vigorously stirred mixture of a diazonium salt and the aromatic substrate, is another useful method for effecting homolytic arylations. The method is much less clean than the decomposition of diaroyl peroxides. A recent modification involves the aprotic diazotisation of an aromatic amine with pentyl nitrite (the pseudo-Gomberg reaction): this results in the formation *in situ* of a diazonium salt thus avoiding the use of a two-phase system. The diazonium salt then reacts further to give aryl radicals. Both these methods have the advantage that they employ readily available aromatic amines. Arylation in a homogeneous system is also possible by use of acylarylnitrosamines, though this procedure suffers from the disadvantage that they have first to be prepared from the amine. Acylarylnitrosamines are now believed to generate aryl radicals as shown in scheme 1:

$$PhN(NO)Ac \rightarrow Ph-N=N-OAc \rightarrow PhN_2^+ + AcO^-$$

$$PhN(NO)Ac + AcO^- \rightarrow Ph-N=N-O^- + Ac_2O$$

$$PhN_2^+ + Ph-N=N-O^- \rightarrow (PhN=N)_2O$$

$$(PhN=N)_2O \rightarrow Ph\cdot + N_2 + Ph-N=N-O\cdot$$

$$Ph\cdot + PhH \rightarrow [Ph.PhH]\cdot$$

$$[Ph.PhH]\cdot + Ph-N=N-O\cdot \rightarrow Ph_2 + Ph-N=N-OH$$

and/or

$$[Ph.PhH]\cdot + PhN_2^+ \rightarrow [Ph.PhH]^+ + Ph\cdot + N_2$$

Scheme 1

A similar mechanism may well be operative in both the Gomberg and pseudo-Gomberg reactions.

12.3. Arylation

12.3.1. *Isomer distributions in reactions of substituted benzenes*

The most notable feature about the phenylation of the benzene derivatives is that all substituents in the benzene nucleus, irrespective of their polar character, exert a small activating influence and that predominant *ortho/para* substitution occurs except when steric effects are important, as in the phenylation of t-butylbenzene (see table 12.1). It must be remembered that conclusions regarding directive effects are based solely on the isomer distribution in the biaryls which are produced. The cyclohexadienyl radicals which are formed initially also dimerize and it is by no means obvious why the formation of the biaryl should account for the same fraction of each of the three isomeric cyclohexadienyl radicals (**4-6**). Nevertheless the proportions of the three isomeric biaryls are remarkably constant even when the arylations are

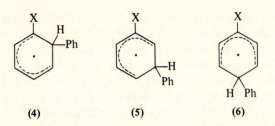

(4)   (5)   (6)

carried out in the presence of oxidizing agents which convert the cyclohexa-dienyl radicals efficiently to the biaryls. It is apparently valid to conclude that both electron-donating and electron-withdrawing substituents do promote *ortho-* and *para*-substitution. This preference for *ortho/para* attack, although it is very much less pronounced than in electrophilic aromatic substitution, arises because both electron-donating and electron-withdrawing substituents can delocalize the unpaired electron onto the substituent when attack occurs *ortho* or *para* to that substituent (cf. **7** and **8**).

Table 12.1. *Relative rates (benzene = 1), isomer ratios and partial rate factors for the phenylation of benzene derivatives (PhR) with dibenzoyl peroxide at 80°C*

R	Relative rate	Isomer distribution (per cent)			Partial rate factors		
		o	*m*	*p*	F_o	F_m	F_p
PhNO$_2$	2.94	62.5	9.8	27.7	5.50	0.86	4.90
PhCl	2.20	59.0	25.0	16.0	3.90	1.65	2.12
PhMe	2.58	60.9	16.1	23.0	4.70	1.24	3.55
PhBu^t	1.09	21.6	50.5	27.9	0.70	1.64	1.81
Ph—Ph	2.94	48.5	23.0	28.5	2.10	1.00	2.50
PhOMe	2.71	55.6	26.9	17.5	2.94	1.42	1.85

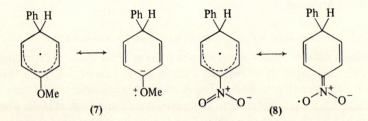

(7)   (8)

The relative rates of phenylation of substituted benzenes were obtained from competition experiments in which a small amount of dibenzoyl peroxide was decomposed in an equimolar mixture of the substituted benzene and benzene (the latter being used as the reference substrate). The relative reactivity of the aromatic substrate $\left(\dfrac{C_6H_5R}{C_6H_6} K \right)$ is equal to the ratio of substituted biphenyls to biphenyl. Substrates other than benzene (e.g. *p*-dichloro-benzene) may also be used as the reference material and the relative reactivity of the substrate, C_6H_5R, obtained using the expression:

$$\frac{C_6H_5R}{C_6H_6}K = \frac{C_6H_5R}{C_6H_4Cl_2}K \times \frac{C_6H_4Cl_2}{C_6H_6}K$$

The very small variation in relative rates obtained in phenylations of substituted benzenes compared with the enormous variation encountered in electrophilic aromatic substitutions such as nitration could make one question as to whether the phenylation reaction is diffusion controlled, in which case little variation in rate with substrate would be expected. However, experimental data indicate that the rate of reaction of phenyl radicals with benzene is about 10^3 times less than would be expected if the reaction rate was diffusion controlled.

The partial rate factor for attack at the *ortho*- position, F_o, is invariably higher than that at the *para*-position, F_p, except when steric effects are important. (The partial rate factors measure the relative reactivity at a particular position in a substituted benzene relative to one position in benzene.) The value of F_m is also invariably greater than one, indicating that *all* positions in a substituted benzene are more reactive than any position in benzene. Even more unusual is the observation that the relative reactivities at the 2-, 4- and 5-positions in the phenylation of *m*-dichlorobenzene are approximately 4 : 2 : 1, i.e. the most reactive position is that between the two substituents. These results are difficult to explain though they do suggest that bond forming is much less advanced in the transition state leading to a cyclohexadienyl radical than in the formation of the Wheland intermediate in electrophilic aromatic substitution. If bond formation is only slightly advanced in the transition state, steric effects will be relatively unimportant. This is consistent with Hammond's postulate since formation of a cyclohexadienyl radical is much less endothermic than formation of a Wheland intermediate.

12.3.2. The influence of polar effects

Arylations using substituted phenyl radicals have indicated that *p*-nitrophenyl radicals are slightly electrophilic and *p*-tolyl radicals are slightly nucleophilic. Thus *p*-nitrophenyl radicals react rather less readily with nitrobenzene than with anisole and give somewhat more of the *meta*-arylated product. The converse is true with *p*-tolyl radicals.

12.3.3. Side-chain attack in reactions of alkylbenzenes

In addition to undergoing arylation in the nucleus, alkylbenzenes also suffer hydrogen abstraction; the resultant benzylic radicals dimerizing:

$$PhCH_3 + Ph\cdot \rightarrow PhCH_2\cdot + PhH$$

$$2PhCH_2\cdot \rightarrow PhCH_2CH_2Ph$$

The extent of side-chain attack increases with increasing stability of the benzylic radical, i.e. with the degree of substitution at the radical centre. The extent of side-chain abstraction in phenylations of alkylbenzenes is in the order: PhBu^t (0 per cent), PhMe (13 per cent), PhEt (55 per cent) and PhPr^i (61 per cent). The amount of side-chain attack is greater with more nucleophilic radicals. A nucleophilic species would be less prone to attack the electron-rich π-system than the methyl group. In agreement with this it has been observed that substitution of the benzene ring with electron-withdrawing substituents decreases the extent of side-chain attack while alkyl radicals react with toluene almost exclusively by hydrogen abstraction from the methyl group.

12.3.4. The effect of additives in arylations with diaroyl peroxides

The yield of biaryl in phenylations with dibenzoyl peroxide is dramatically increased by carrying out the reactions in presence of certain additives, notably oxygen, copper(II) benzoate or nitro-compounds (see table 12.2).

Table 12.2. *Effect of additives on the yield of biaryl produced in the decomposition of diaroyl peroxides in aromatic substrates*

Peroxide	Substrate	Additive	Yield (per cent)
$(PhCO_2)_2$	PhH	–	40
$(PhCO_2)_2$	PhH	$m\text{-}C_6H_4(NO_2)_2$	85
$(PhCO_2)_2$	PhH	$Cu(OCOPh)_2$	88
$(PhCO_2)_2$	PhH	$Fe(OCOPh)_3$	90
$(PhCO_2)_2$	PhH	O_2	$126-151^a$

[a]Based on one mole of biaryl being produced from one mole of peroxide.

Oxygen acts by abstracting hydrogen with the generation of a hydroperoxy radical. More convenient is the use of copper(II) benzoate which oxidizes the phenylcyclohexadienyl radicals to the corresponding cation. The resultant copper(I) benzoate is oxidized to copper(II) benzoate by reaction with the peroxide.

$$[Ph.PhH] \cdot + Cu^{2+} \rightarrow [Ph.PhH]^+ + Cu^+$$

$$[Ph.PhH]^+ \rightarrow Ph_2 + H^+$$

$$(PhCO_2)_2 + Cu^+ \rightarrow PhCO_2Cu^{II} + PhCO_2 \cdot$$

Experimentally this is very satisfactory as the copper(II) benzoate and benzoic acid are readily extracted by sodium hydroxide at the end of the reaction. The mode of action of nitro-compounds is less certain. They may be reduced to the corresponding nitroso-compound which could scavenge phenyl radicals to generate a nitroxide. This latter could dehydrogenate the phenylcyclohexadienyl radicals.

$$PhNO_2 \rightarrow PhNO$$

$$PhNO + Ph \cdot \rightarrow Ph_2 NO \cdot$$

$$[Ph.PhH] \cdot + Ph_2 NO \cdot \rightarrow Ph_2 + Ph_2 NOH$$

$$Ph_2 NOH + (PhCO_2)_2 \rightarrow Ph_2 NO \cdot + PhCO_2 H + PhCO_2 \cdot$$

Alternatively they may act as electron-transfer oxidants.

$$[Ph.PhH] \cdot + PhNO_2 \rightarrow [Ph.PhH]^+ + PhNO_2 \overline{\cdot}$$

$$[Ph.PhH]^+ \rightarrow Ph_2 + H^+$$

$$PhNO_2 \overline{\cdot} + (PhCO_2)_2 \rightarrow PhNO_2 + PhCO_2^- + PhCO_2 \cdot$$

12.3.5. *Intramolecular arylations*

The Pschorr reaction, which involves the copper-catalysed cyclization of the diazonium salt of *trans*-2-amino-α-phenylcinnamic acid to phenanthrene-9-carboxylic acid, involves homolytic intramolecular arylation (scheme 2). The

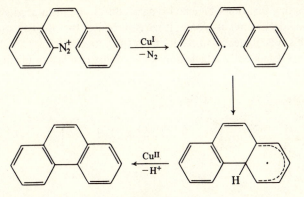

Scheme 2

copper effects reduction of the diazonium salt to the aryl radical with concomitant loss of nitrogen (see p. 146). Cyclization of the intermediate radical occurs in preference to reaction with solvent because of the close proximity of the radical centre to the second aromatic system. Intramolecular radical cyclization to give a five-membered ring as should occur in the copper-catalysed decomposition of the diazonium salt of *o*-aminobenzophenone (**9**)

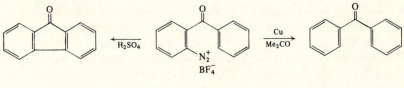

(9)

occurs very much less readily; the main reaction product being benzophenone formed as a result of hydrogen abstraction from the solvent by the inter-mediate radical. Heating this diazonium salt in sulphuric acid, however, gives fluorenone as a result of cyclization of the intermediate cation. It appears that cyclization by a heterolytic route is less critically controlled by steric factors than is radical cyclization.

12.4. Acyloxylation

In arylations with diaroyl peroxides, acyloxylation is a competing reaction. It only occurs to an appreciable extent with substrates such as anisole and poly-cyclic aromatic hydrocarbons. Thus reaction of dibenzoyl peroxide with naphthalene gives α- and β-benzoyloxynaphthalenes in addition to α- and β-phenylnaphthalenes.

The extent of benzoyloxylation of anisole and other methoxybenzenes is far greater than would be expected on the basis of the activating influence of the methoxy group. Thus the reaction of dibenzoyl peroxide with *p*-methyl-anisole gave neither biaryl nor bibenzyl-type products: the main product was 2-methoxy-5-methylphenyl benzoate. It seems probable that the key step in reactions of this type is electron transfer from the peroxide to give a radical cation which is subsequently attacked by a benzoyloxy radical or dibenzoyl peroxide (see scheme 3).

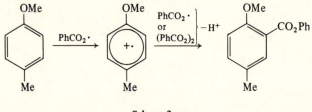

Scheme 3

12.5. Alkylation

Homolytic alkylation of aromatic compounds is a much less valuable syn-thetic reaction than arylation due to the prevalence of competing reactions. A comparison of the isomer ratios obtained in the methylation of monosubsti-tuted benzenes with those for phenylation indicate that methyl radicals are more nucleophilic than phenyl radicals. Thus the *meta : para* ratios for the methylation of monosubstituted benzenes containing electron-donating substituents are greater than those for phenylation of the same compounds. (Results on substitution at the *ortho*-position have not been used because of steric effects). The reverse is true for substrates containing electron-withdraw-ing substituents. These trends are even more pronounced in cyclohexylations, whereas in phenylethynylations the *meta : para* ratios indicate that phenyl-ethynyl radicals, $PhC\equiv C \cdot$ are electrophilic.

Hammett plots for F_m and F_p for both the cyclohexylation and phenyl-ethynylation reactions give ρ values of +1.10 for cyclohexylation and −1.56 for phenylethynylation. The ρ value for phenylations is +0.05. (A negative value of the Hammett reaction constant, ρ, indicates the reaction is facilitated by electron-withdrawing substituents in the reactant, whilst a positive ρ value means the reaction is accelerated by electron-donating substituents. For a further discussion of the Hammett equation and its applications to the study of organic reaction mechanisms, see C. D. Johnson, *The Hammett Equation*, Cambridge University Press, 1973.)

12.6. Hydroxylation

The radical hydroxylation of aromatic compounds, though not a reaction of synthetic utility, is of considerable importance from a biological point of view. A 'foreign' aromatic compound is usually hydroxylated when admin-istered to animals. Thus, benzoic acid is converted into a mixture of *ortho–meta*-, and *para*-hydroxybenzoic acids. The formation of all three isomers led to the belief that hydroxylation proceeded by a radical process. Consequently, it is of considerable interest to study the hydroxylation of aromatic com-pounds in orthodox organic chemistry and to compare the products with those from biological hydroxylations.

Hydroxyl radicals, generated in Fenton's reaction (see p. 145), add to benzene to give initially hydroxycyclohexadienyl radicals which have been identified from their ESR spectra.

$$HO \cdot + PhH \rightarrow [HO.PhH] \cdot$$

The hydroxycyclohexadienyl radicals subsequently dimerize yielding, after dehydration of the dimer, biphenyl. Alternatively they are oxidized to give phenol: this route is favoured in the presence of added oxidants such as copper(II) salts, iron(III) salts or oxygen.

$$[HO.PhH] \cdot \begin{array}{c} \nearrow \quad [HO.PhH]_2 \xrightarrow{-2H_2O} Ph_2 \\[2mm] \searrow_{Cu^{2+}} \quad [HO.PhH]^+ \xrightarrow{H^+} PhOH \end{array}$$

The hydroxyl radical has unmistakable electrophilic characteristics as shown by the relative rates of reactivity and isomer distributions obtained from a series of monosubstituted benzenes. Thus electron-withdrawing groups decrease the rate of reaction and at the same time increase the proportion of the *meta*-hydroxylated product (see table 12.3).

Table 12.3. *Hydroxylation of aromatic compounds*

Aromatic compound, PhX	Relative reactivity $\frac{PhX}{PhH} K$	Isomer distribution (per cent)		
		o	*m*	*p*
PhOMe	6.35	85	4	11
PhMe	–	55	15	30
PhF	–	37	18	45
PhCl	0.55	42	29	29
PhNO$_2$	0.14	24	30	46

12.7. Homolytic substitutions of polycyclic aromatic hydrocarbons

Monocyclic aromatic compounds are not in general attacked by the relatively stable benzyl radicals nor by acyloxy radicals (these latter undergo decarboxylation rather than reaction with the aromatic substrate). The formation of the adduct radicals from polycyclic aromatic compounds is much more exothermic than the formation of the analogous radicals from benzene. Thus anthracene undergoes both benzylation and benzoyloxylation. In the reaction of anthracene with benzyl radicals the products include both 9-benzylanthracene and 9,10-dibenzyl-9,10-dihydroanthracene: these arise from reaction of benzyl radicals with the intermediate σ-radicals (10) (scheme 4). In addition these radicals also dimerize to give 10,10'-dibenzyl-9,9',10,10'-tetrahydro 9,9'-dianthryl.

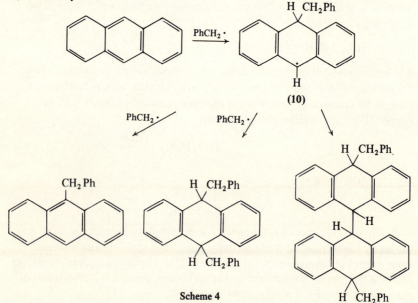

Scheme 4

12.8. Homolytic substitutions of heteroaromatic compounds

In general heteroaromatic compounds undergo reactions with radicals in much the same way as their carbocyclic analogues. Attack tends to take place initially at a carbon atom rather than at the heteroatom. Thus reaction of benzyl radicals with acridine occurs at the 10-position: the resultant radical then reacts further with benzyl radicals to give 10-benzylacridine and 5,10-dibenzylacridan (scheme 5).

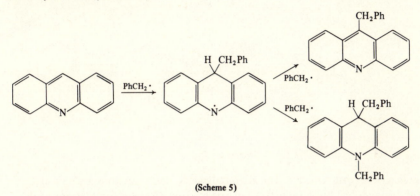

(Scheme 5)

In no instance are products derived from N—N coupling isolated: this is attributed to the ease with which such compounds would undergo homolysis.

Yields of products derived from alkylation of heterocyclic compounds are generally poor: in addition selectivity is low. A very different situation prevails in the alkylation of protonated heterocyclic compounds including pyridines, quinolines, isoquinolines, acridine, quinoxaline, pyrimidines, thiazoles and imidazoles. Under these conditions the yields of alkylated products are much higher and reaction occurs exclusively α and γ to the protonated nitrogen (reaction 8).

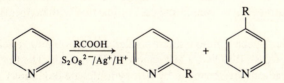

$$\tag{8}$$

Quantitative studies on the rates of alkylation of 4-substituted pyridines are consistent with alkyl radicals possessing appreciable nucleophilic character, i.e. the reactions are facilitated by electron-donating substituents. The high sensitivity to polar effects in this class of reaction has been interpreted in terms of a polar π-complex (**11**) preceding the formation of the σ-complexes (**12** and **13**) (scheme 6). The order of nucleophilicity of alkyl radicals is: Me· < Pr^i· < Bu^t· < PhCH$_2$·; alkylations with benzyl radicals being the most sensitive to polar influences in the substrate and those with methyl

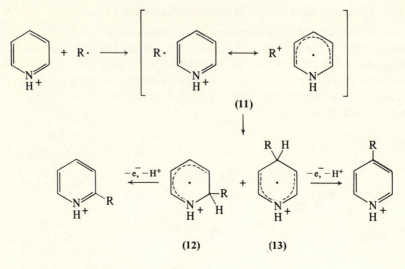

Scheme 6

radicals the least sensitive. The alkyl radicals in these reactions are conveniently generated by the silver(I)-catalysed peroxydisulphate oxidations of the appropriate carboxylic acid.

$$S_2O_8{}^{2-} + Ag^+ \to SO_4{}^{\overline{\cdot}} + SO_4{}^{2-} + Ag^{2+}$$

$$RCOOH + Ag^{2+} \to R\cdot + CO_2 + Ag^+$$

This system also oxidizes the σ-radical to the alkylated product.

Synthetically the method is valuable in that it allows the preparation of alkylated heterocyclic compounds not otherwise readily synthesized. Moreover it is applicable for the introduction of t-butyl and benzyl groups which cannot be introduced by conventional homolytic alkylation. Substitution occurs without rearrangement even in the case of reaction with neopentyl radicals.

Protonated heterocyclic compounds are similarly susceptible to attack by $R\dot{C}O$, $R_2\dot{C}OH$, $R_2\dot{C}OR'$, $R'CONR^2\dot{C}HR^3$ and $\dot{C}ONR_2$ radicals. The first four of these types of radicals are generated by peroxydisulphate oxidations of α-keto acids, alcohols, ethers and N-alkylamides respectively whereas carbamoyl radicals, $\dot{C}ONR_2$, are generated by oxidation of N-substituted formamides with t-butyl hydroperoxide and iron(II). All these reactions show the same characteristics as the alkylation reaction, i.e. the reactions are facilitated by electron-donating substituents in the heterocyclic compound which is consistent with the attacking radicals having nucleophilic character.

13   Radical oxidations and reductions

13.1.   Introduction

The distinctive feature of homolytic oxidations and reductions is that the
oxidation or reduction step involves the transfer of a single electron or its
equivalent (reactions *1* and *2*) rather than two electrons (reaction *3*).

$$ArO^- + Ce^{4+} \rightarrow ArO\cdot + Ce^{3+} \tag{1}$$

$$Ar\cdot + CuCl_2 \rightarrow ArCl + CuCl \tag{2}$$

$$R_2\overset{\text{H}}{\underset{}{C}}-O-CrO_3H + B: \longrightarrow R_2CO + HCrO_3^- + BH^+ \tag{3}$$

Two distinct steps can be delineated for metal-catalysed oxidations and
reductions and the analogous electrochemical reactions:
(1) oxidation or reduction of the organic compound to give a radical (or
    radical ion);
(2) subsequent reaction of the radical, which may involve its further oxida-
    tion or reduction. Alternatively radical–radical reactions may occur.
These steps can be outlined by reference to the electrochemical oxidation of
carboxylic acids (the Kolbe reaction):

$$RCOO^- - e^- \rightarrow RCOO\cdot \rightarrow R\cdot + CO_2$$

$$R\cdot - e^- \rightarrow R^+ \rightarrow Products$$

$$R\cdot + R\cdot \rightarrow R-R$$

A large number of metal oxidants, e.g. Ag^+, Ce^{4+}, Co^{3+}, Cu^{2+}, $Fe(CN)_6^{3-}$,
$IrCl_6^{2-}$, Mn^{3+}, MnO_4^- and VO_2^+, are single-electron oxidants and hence will
give rise to radical intermediates. The examples chosen in this chapter have been
selected so as to illustrate reaction mechanisms though many of the processes
discussed have significant synthetic potential.

13.2. Oxidation of radicals

There are basically two alternative routes whereby radicals may be oxidized, namely electron transfer (reaction *4*) and ligand transfer (reaction *5*). These are analogous to outer-sphere and inner-sphere oxidations of metal complexes.

$$R\cdot + Cu^{2+} \rightarrow R^+ + Cu^+ \tag{4}$$

$$R\cdot + CuCl_2 \rightarrow RCl + CuCl \tag{5}$$

13.2.1. Electron-transfer oxidation

This essentially involves the oxidation of a radical to a carbonium ion which subsequently reacts to give stable products. The product composition can be used to provide evidence for the intermediacy of carbonium ions. Thus oxidation of cyclobutyl radicals by lead(IV) gives the same mixture of acetates as that obtained in solvolyses proceeding via the carbonium ion reactions (*6* and *7*).

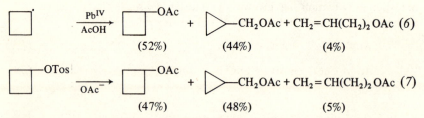

$$\text{(52\%)} \qquad \text{(44\%)} \qquad \text{(4\%)} \tag{6}$$

$$\text{(47\%)} \qquad \text{(48\%)} \qquad \text{(5\%)} \tag{7}$$

Cyclobutyl radicals unlike cyclobutyl cations do not undergo rearrangement.

The ease of oxidation of a radical to a carbonium ion will depend on the oxidation potential of both the radical and the oxidant. Thus, *p*-methoxybenzyl radicals are much more readily oxidized by lead(IV) than benzyl radicals, owing to the stabilizing effect the methoxy group has on the *p*-methoxybenzyl cation (the methoxy group has minimal stabilizing effect on the radical) indicating that there is considerable development of carbonium-ion character in the transition state in electron-transfer oxidation of radicals.

Copper(II) is a very much more efficient oxidant than would be expected on the basis of its oxidation potential. The products from oxidations with copper(II) are largely alkenes rather than products derived from carbonium ions. Thus oxidation of cyclobutyl radicals with copper(II) acetate gives almost exclusively cyclobutene. To account for this it has been proposed that these reactions involve an intermediate alkylcopper species which undergoes a *cis*-elimination to give the alkene. Use has been made of this reaction in the

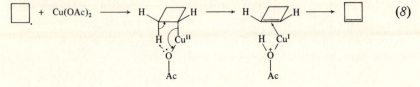

synthesis of alkenes from the lead(IV) acetate decarboxylation of carboxylic acids in presence of copper(II) acetate:

$$RCH_2CH_2CO_2H \xrightarrow{Pb(OAc)_4} RCH_2CH_2 \cdot \xrightarrow{Cu(OAc)_2} RCH=CH_2$$

13.2.2. Ligand-transfer oxidation

Ligand-transfer oxidations involve the direct transfer of a group from the metal salt to the radical. The most commonly encountered examples involve the transfer of halogen but other groups such as azide, thiocyanate and xanthate can also be involved.

An important distinction between ligand- and electron-transfer processes is that the former but not the latter is relatively insensitive to electronic effects in the radical, i.e. there is only a small degree of electron transfer in the transition state for the reaction, which can be represented as:

$$[R \cdot Cl-Cu^{II}-Cl \leftrightarrow R-Cl \quad Cu^{I}-Cl]$$

The transition state is thus more akin to that of an atom transfer in a typical free-radical reaction. As there is little charge development in the transition state in ligand-transfer reactions, it is found that radicals with electron-withdrawing substituents are sometimes oxidized by ligand-transfer but not by electron-transfer oxidants (see reaction 9).

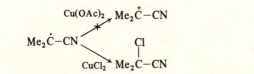

$$(9)$$

Another important feature of ligand-transfer processes is that they proceed without rearrangement of the alkyl group. Thus cyclobutyl chloride is formed in the reaction of cyclobutyl radicals with copper(II) chloride (reaction 10):

$$\text{cyclobutyl} \cdot \xrightarrow{CuCl_2} \text{cyclobutyl-Cl} \qquad (10)$$

no cyclopropylmethyl chloride or 1-chlorobut-3-ene are obtained (cf. reaction 6 for the corresponding electron-transfer reaction).

Alkyl halides can be synthesized from carboxylic acid by reaction with lead(IV) acetate and the appropriate lithium halide: the intermediate alkyl radicals undergo ligand-transfer oxidation with a halo-lead(IV) complex.

$$RCO_2H \xrightarrow{Pb(OAc)_4} R \cdot \xrightarrow{XPb^{(IV)}} RX$$

13.3.   Reduction of radicals

Both electron- and ligand-transfer processes can also be described for the reduction of organic radicals. There are fewer examples of reactions involving the reduction of radicals, reflecting the fact that radicals are rather more readily oxidized to carbonium ions than reduced to carbanions. This is consistent with the nucleophilic character which alkyl radicals display in homolytic aromatic substitution (see p. 130). Radicals containing electron-withdrawing substituents such as α-ketoalkyl radicals are reduced relatively readily (see p. 146).

$$CH_3COCH_2 \cdot \overset{+e^-}{\rightarrow} CH_3COCH_2^-$$

13.4.   Mechanism for oxidations of organic compounds with metal-ion oxidants

13.4.1.   Oxidations of 1,2-glycols

Kinetic studies have established that oxidation of pinacol and other di-tertiary glycols with vanadium(V) proceeds via a cyclic chelate (reaction *11*). The reaction leads to ketone plus an organic free radical: this latter can be detected by its ability to promote polymerization of vinyl monomers. Two-electron oxidants, e.g. lead(IV), in contrast give directly two molecules of ketone (reaction *12*).

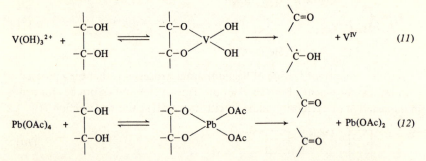

Pinacol is resistant to oxidation by hexachloroiridate and iron(III) tris-*o*-phenanthroline, both of which effect oxidation by simple electron transfer to the substrate. The resistance to oxidation of pinacol can be attributed to the absence of low-energy outer-sphere oxidation routes (reaction *13*): the resultant radical cation would be of higher energy. This is in spite of hexachloro-iridate being, in terms of its oxidation potential, a stronger oxidant than vanadium(V).

$$\begin{array}{c}
\text{Me}_2\text{C--OH} \\
| \\
\text{Me}_2\text{C--OH}
\end{array}
+ \text{IrCl}_6{}^{2-} \overset{\times}{\longrightarrow}
\begin{array}{c}
\text{Me}_2\overset{+\cdot}{\text{C}}\text{--OH} \\
| \\
\text{Me}_2\text{C--OH}
\end{array}
+ \text{IrCl}_6{}^{3-} \qquad (13)$$

13.4.2. Oxidations of aromatic compounds

Aromatic compounds are susceptible to oxidation by one-electron oxidants by an outer-sphere mechanism to give initially a radical cation: this can in certain instances be detected by ESR spectroscopy. Whether or not oxidation occurs depends on the oxidation potentials of the substrate and the oxidant.

$$\text{ArH} \xrightarrow{-e^-} \text{ArH}^{+\bullet}$$

Thus *p*-methoxytoluene but not toluene is oxidized by manganese(III) acetate, whereas both compounds are oxidized by a stronger oxidant such as cobalt(III) acetate.

The formation of the radical cation can also be achieved electrochemically. Electrodes act as electron sinks in anodic oxidations. The ability of a particular organic compound to undergo oxidation depends on the potential at which the process is carried out. The radical cation generated either chemically or electrochemically can then undergo nucleophilic capture by the solvent. Thus anodic oxidation of biphenyl in the presence of acetate gives a mixture of *o*- and *p*-acetoxybiphenyls: these result from attack of acetate ions on the biphenyl radical cation (scheme 1). This process occurs at a potential below that at which acetate ions are discharged, indicating that the products do not arise by attack of acetoxy radicals on the biphenyl. This type of process is

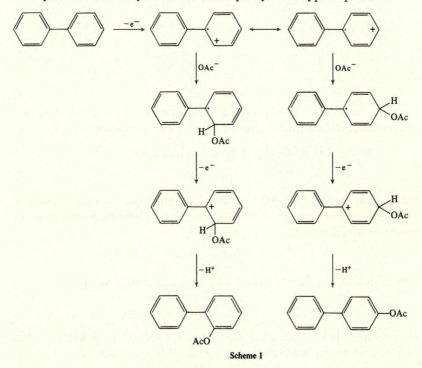

Scheme 1

referred to as an ECE process (electrochemical–chemical–electrochemical) indicating that the initial electron transfer is followed by a chemical step and this in turn by a second electron transfer.

The radical cations generated in oxidations of aromatic compounds can react with a molecule of this substrate to give dimeric radical cations and thence biaryl. This route is more important in poorly nucleophilic solvents such as trifluoroacetic acid. In oxidations of methylbenzenes, the derived radical cations easily lose a proton to give benzylic radicals and thence by further oxidation benzylic cations. These latter are either captured by the solvent or they attack another molecule of substrate to give diarylmethanes (scheme 2).

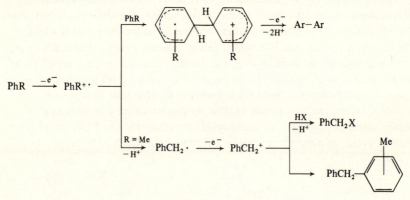

Scheme 2

13.5. Kolbe oxidation of carboxylic acids

The Kolbe reaction involving the anodic oxidation of carboxylates is a useful synthetic procedure for the synthesis of long-chain compounds.

$$MeO_2C(CH_2)_8CO_2H \rightarrow MeO_2C(CH_2)_{16}CO_2Me$$

$$(69-74\%)$$

The reaction involves the C–C coupling of two alkyl radicals. Products derived from the disproportionation of the alkyl radicals and from their reaction with solvent are also observed.

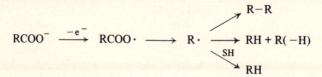

Alcohols and esters are frequently isolated as by-products in the Kolbe reaction: these result from further oxidation of the radical to the cation:

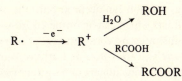

The proportion of products resulting from carbonium ions is greater the more easily oxidized is the intermediate radical. Thus bibenzyl is the major product from the oxidation of phenylacetic acid, while *p*-methoxyphenylacetic acid gives only 1 per cent of 4,4′-dimethoxybibenzyl and 99 per cent of *p*-methoxybenzyl methyl ether resulting from capture of the *p*-methoxybenzyl cations by the solvent methanol.

$$ArCH_2CO_2^- \xrightarrow{-e^-} ArCH_2CO_2 \cdot \longrightarrow ArCH_2 \cdot \xrightarrow{-e^-} ArCH_2^+$$

$$\downarrow \qquad\qquad\qquad\qquad \downarrow MeOH/-H^+$$

$$ArCH_2 CH_2 Ar \qquad ArCH_2 OMe$$

The relative amount of products resulting from the carbonium ion route is considerably greater when oxidations are carried out with carbon rather than platinum electrodes.

13.6. Oxidation of phenols

The oxidation of phenols is a topic of immense importance because of the widespread occurrence of phenolic coupling in the biosynthesis of natural products and also because of the role of phenols as antioxidants.

Phenols can be oxidized by a wide variety of one-electron oxidants, including enzyme systems, and also by electrochemical methods. Oxidation of either the phenol molecule or its anion occurs to give phenoxy radicals (these can be detected by ESR spectroscopy). The resultant phenoxy radicals can then couple to give a number of isomeric dimeric products. This seems to be the most probable mechanism in oxidations of simple phenols by metal oxidants.

$$ArOH \xrightarrow{-e^-} Ar\overset{+\cdot}{O}H \xrightarrow{-H^+} ArO \cdot$$

$$ArO^- \xrightarrow{-e^-} ArO \cdot$$

$$2ArO \cdot \rightarrow \text{dimeric products}$$

Coupling of phenoxy radicals, in which the unpaired spin is delocalized onto the *ortho*- and *para*-positions of the benzene ring (see p. 26), produces dimers which are formed by C—C and C—O but not O—O combination. The latter does not occur on account of the instability of the resultant peroxide.

In the case of C—C and C—O coupling, reaction can occur at both the *ortho*- and *para*-positions. The variety of products which can be obtained from the oxidative coupling of phenols is well illustrated by consideration of *p*-cresol. Here products are derived from *ortho–ortho*, *ortho–para* and *ortho*-O coupling (see scheme 3). The initial cyclohexadienone dimers undergo enolization to give the biphenol (**4**), the phenolic ether (**6**), and, after an intramolecular Michael-type addition, Pummerer's ketone (**5**). No products are derived from *para*-O and *para–para* coupling: enolization of these coupled products could not occur and such dimers would dissociate back to *p*-methylphenoxy radicals. Kinetic and product studies indicate that the initial coupling step to give the cyclohexadienone dimers is reversible. No product analogous to Pummerer's ketone is obtained in the oxidation of *p*-t-butylphenol: steric effects reduce the stability of the cyclohexadienone dimer analogous to (**2**) which thus dissociates.

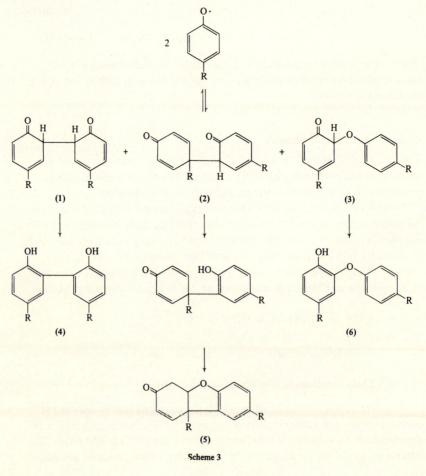

Scheme 3

The same coupled products could also be formed by two other mechanisms involving radical substitution of a phenol molecule followed by subsequent oxidation of the intermediate cyclohexadienyl radical, or by oxidation of the phenoxy radical to the corresponding phenoxonium ion followed by electrophilic substitution by this on a phenol molecule. These processes are illustrated in reactions *(14)* and *(15)* for the formation of the C—O coupled products (similar reactions could be formulated for the formation of the other dimeric products). There is evidence that in certain circumstances both these mechanisms are operative. The radical substitution pathway is only important when intramolecular attack can occur.

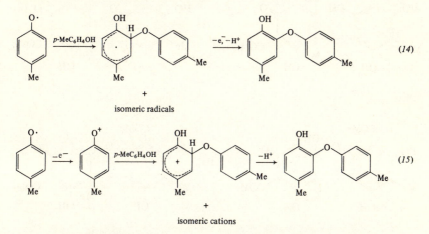

A crucial step in the biosynthesis of many alkaloids and other natural products including lignins, aphid pigments and antibiotics, is generally recognized as involving phenolic oxidation by enzymes acting as one-electron oxidants. Evidence for the involvement of this mechanism is largely based on experiments involving the use of labelled precursors. When analogous oxidations have been carried out *in vitro* the yields of coupled product are much lower than those obtained *in vivo*. The enzymes in the plants or animals fold the molecules undergoing oxidation such that only one coupled product is obtained. One of the best known examples of oxidative coupling in biosynthesis relates to the formation of usnic acid which is structurally closely related to Pummerer's ketone. The synthesis of usnic acid has been achieved by the ferricyanide oxidation of *C*-methylphloroacetophenone followed by dehydration of the coupled product (scheme 4).

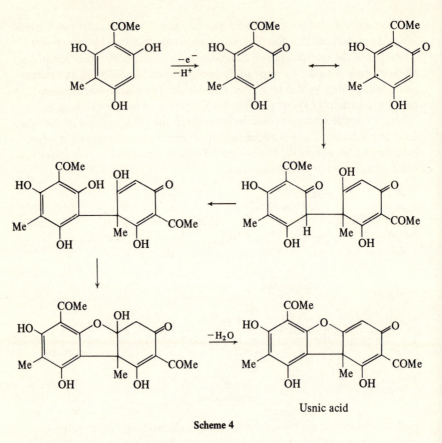

Usnic acid

Scheme 4

13.7. Metal-catalysed reductions of organic compounds

Organic compounds can undergo homolytic reduction by single-electron transfer in much the same way as they undergo oxidation, though there are fewer examples known than of oxidation. One of the oldest and most useful reduction procedures is the dissolving-metal reduction or, as it is commonly called, Birch reduction. A solution of a strongly electropositive metal, e.g. sodium in an inert solvent, commonly liquid ammonia or a low molecular weight amine, containing a proton source (this is often an alcohol) is employed for the reduction. The metal acts as an electron donor and can bring about the reduction of a double bond to the resonance-stabilized radical anion.

$$X{=}Y + M \rightarrow [\overset{..}{\overset{..}{X}}{-}\overset{.}{Y} \leftrightarrow \overset{.}{X}{-}\overset{..}{\overset{..}{Y}}] + M^+$$

The radical anion is then protonated to give a radical.

$$[\overset{..}{\overset{..}{X}}{-}\overset{.}{Y} \leftrightarrow \overset{.}{X}{-}\overset{..}{\overset{..}{Y}}] + H^+ \rightarrow \overset{.}{X}{-}YH$$

Further reduction and proton capture gives the reduced compound.

$$\dot{X}-YH + M \rightarrow \bar{\bar{X}}-YH + M^+$$

$$\bar{\bar{X}}-YH + H^+ \rightarrow HX-YH$$

This procedure is widely used in the reduction of aromatic compounds. Thus benzene is reduced by lithium in liquid ammonia containing a little ethanol to dihydrobenzene (reaction *16*).

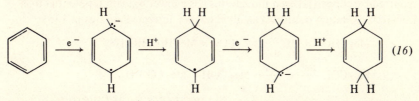

(16)

Reduction of monosubstituted benzenes could lead to either the 1-substituted 1,4-dihydrobenzene or the 1-substituted 2,5-dihydrobenzene. The former is obtained for benzenes containing electron-withdrawing substituents and the latter if the substituent is electron-donating: in each case reaction occurs via the more stable radical anion.

Dissolving metal reductions can similarly be used for the reduction of alkynes to *trans*-alkenes (reaction *17*) and of ketones to alcohols (reaction *18*).

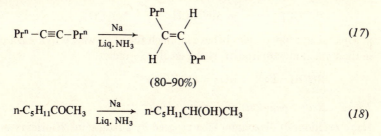

(17)

(80–90%)

$$n\text{-}C_5H_{11}COCH_3 \xrightarrow[\text{Liq. NH}_3]{\text{Na}} n\text{-}C_5H_{11}CH(OH)CH_3 \qquad (18)$$

13.8. Redox reactions

13.8.1. *Fenton's reaction*

The generation of hydroxyl radicals in the reaction of hydrogen peroxide with iron(II) sulphate (Fenton's reaction) was one of the first radical reactions discovered.

$$Fe^{2+} + H_2O_2 \rightarrow Fe^{3+} + HO^- + HO\cdot$$

The hydroxyl radicals react with an organic substrate either by hydrogen abstraction or by addition to an unsaturated system.

$$RH + HO\cdot \longrightarrow R\cdot + H_2O$$

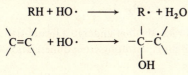

The organic radicals thus generated may dimerize, be further oxidized by the iron(III) generated or by added copper(II), or be reduced by iron(II). The preferred reaction pathway depends on the redox potential of the radical and also on the reaction conditions. Thus while $Me_2\overset{\bullet}{C}OH$ radicals are readily oxidized by iron(III), the isomeric $MeCH(OH)CH_2\cdot$ radicals, which would have a higher oxidation potential, dimerize. Both radicals, are, however, oxidized by copper(II) even though this has a lower oxidation potential than iron(III): oxidation occurs via an alkyl copper intermediate (see p. 136).

$$Me_2CHOH + HO\cdot \rightarrow Me_2\overset{\bullet}{C}OH + MeCH(OH)CH_2\cdot$$

$$Me_2\overset{\bullet}{C}OH \xrightarrow{\text{Fe}^{III} \text{ or } \text{Cu}^{II}} Me_2\overset{+}{C}OH \rightarrow Me_2CO + H^+$$

$$MeCH(OH)CH_2\cdot \begin{cases} \longrightarrow MeCH(OH)CH_2CH_2CH(OH)Me \\ \xrightarrow{\text{Cu}^{II}OH} MeCH(OH)CH_2OH \end{cases}$$

Reduction becomes important with radicals containing electron-withdrawing groups which can delocalize the charge in the resultant carbanion.

$$MeCOCH_2\cdot \xrightarrow{\text{Fe}^{II}} MeCOCH_2^- \xrightarrow{H^+} Me_2CO$$

Organic hydroperoxides behave in much the same way as hydrogen peroxide in their reaction with iron(II) giving an alkoxy radical.

$$ROOH + Fe^{2+} \rightarrow RO\cdot + HO^- + Fe^{3+}$$

13.8.2. Redox reactions of diazonium salts

Many reactions of diazonium salts proceed by a radical mechanism (see p. 124). This is particularly true of those reactions which are catalysed by copper(I) salts or copper including the Sandmeyer, Gattermann and Meerwein reactions as well as the Pschorr reaction (p. 129).

$$ArN_2Cl \xrightarrow{\text{CuCl}} ArCl + N_2$$

$$ArN_2Cl + CH_2{=}CHX \xrightarrow[\text{Me}_2\text{CO}]{\text{CuCl}_2} ArCH_2CHClX$$

The key step in all these reactions is the electron-transfer reduction of the diazonium salt with the resultant generation of aryl radicals.

$$ArN_2^+ + CuCl_2^- \rightarrow Ar\cdot + N_2 + CuCl_2$$

In the Sandmeyer reaction the aryl radicals then undergo a ligand-transfer reaction with copper(II) chloride. Biaryls, resulting from dimerization of aryl radicals are by-products.

$$Ar \cdot + CuCl_2 \rightarrow ArCl + CuCl$$

$$Ar \cdot + Ar \cdot \rightarrow Ar-Ar$$

As Waters has pointed out, the salient feature of this mechanism is that it can only occur within closely defined redox potential limits. The redox potentials of the Cu^+/Cu^{2+} and Cu/Cu^+ systems in aqueous solution allow for both facile oxidation and reduction as envisaged in the two steps of the Sandmeyer and Gattermann reactions.

The substitution of the diazonium group by hydrogen, a reaction of considerable synthetic utility, is also a radical-chain process involving the participation of aryl radicals.

$$ArN_2^+ + H_2PO_2^- \xrightarrow{\text{slow}} Ar \cdot + N_2 + H_2PO_2 \cdot \qquad (\textit{initiation})$$

$$Ar \cdot + H_3PO_2 \rightarrow ArH + H_2PO_2 \cdot$$

$$ArN_2^+ + H_2PO_2 \cdot \rightarrow Ar \cdot + N_2 + H_2PO_2^+$$

$$H_2PO_2^+ + H_2O \rightarrow H_3PO_3 + H^+$$

Not unexpectedly this reaction is also catalysed by copper(I) salts which initiate the chain reaction rather more rapidly than hypophosphorous acid.

13.9. Electron-transfer substitution reactions

13.9.1. *Of aliphatic compounds*

There is considerable evidence that certain nucleophilic aliphatic substitutions proceed by an electron-transfer mechanism. An excellent example of this type of mechanism is seen in the reaction between *p*-nitrobenzyl chloride and the anion of 2-nitropropane which gives 92 per cent of the *C*-alkylated product.

$$p\text{-}NO_2C_6H_4CH_2Cl + Me_2\bar{C}NO_2 \rightarrow p\text{-}NO_2C_6H_4CH_2CMe_2NO_2 + Cl^-$$

In contrast the corresponding reaction of benzyl chloride is much slower and results in exclusive *O*-alkylation:

$$PhCH_2Cl + Me_2\bar{C}NO_2 \longrightarrow Me_2C=\overset{+}{N}\overset{\diagup O^-}{\underset{\diagdown OCH_2Ph}{}} + Cl^-$$

The reaction of *p*-nitrobenzyl chloride but not that of benzyl chloride is retarded by radical scavengers such as oxygen and di-t-butyl nitroxide implicating a radical-chain mechanism for *C*-alkylation. The reaction is also retarded by *p*-dinitrobenzene but catalysed by electron-transfer agents such as sodium naphthalenide (Na^+, $C_{10}H_8^{\bar{\cdot}}$) suggesting an electron-transfer process (scheme 5). The initial step involves formation of a radical anion which can undergo fragmentation to give the *p*-nitrobenzyl radical and chloride. This step is suppressed by *p*-dinitrobenzene which is a good electron acceptor. The *p*-nitrobenzyl radical then couples with the 2-nitropropane anion to give the radical anion (7), which reacts with *p*-nitrobenzyl chloride in an electron-transfer process to give the product. When this reaction is carried out in the

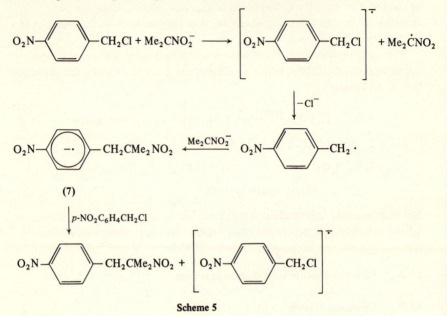

Scheme 5

presence of *p*-dinitrobenzene, significant amounts of *O*-alkylated product, derived by a normal nucleophilic displacement reaction, are obtained.

13.9.2. *Of aromatic compounds*

The nucleophilic substitution of aromatic halides by a radical pathway can be promoted by electron donors, e.g. potassium metal (reduction is probably effected by solvated electrons, e_{NH_3}). In the absence of potassium the reaction

$$ArX \xrightarrow[NH_3]{KNH_2,\ K} ArNH_2$$

proceeds by an aryne mechanism. The formation of *m*-anisidine from the reaction of *o*-haloanisoles with potassium amide epitomizes this latter mechanism (reaction *19*).

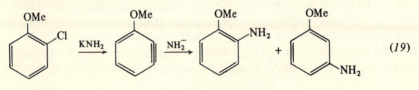

In the presence of potassium metal as an electron donor, *o*-iodoanisole affords *o*-anisidine together with some anisole: no *m*-anisidine is obtained. This reaction is recognized as proceeding via a radical-chain mechanism involving the formation of the radical anion of the haloarene.

$$ArX + e^- \rightarrow ArX^{\cdot -}$$

$$ArX^{\cdot -} \rightarrow Ar\cdot + X^-$$

$$Ar\cdot + NH_2^- \rightarrow ArNH_2^{\cdot -}$$

$$ArNH_2^{\cdot -} + ArX \rightarrow ArNH_2 + ArX^{\cdot -}$$

$$Ar\cdot + NH_3 \rightarrow ArH + NH_2\cdot$$

A similar mechanism operates in the reactions of aryl halides with the enolate anions of carbonyl compounds, and with the anions of acidic hydrocarbons such as fluorene.

$$ArCl + CH_3COCH_2^- \xrightarrow[NH_3]{K} ArCH_2COCH_3$$

That these reactions can also be photostimulated and are retarded by oxygen and di-t-butyl nitroxide, substantiates the radical-chain nature of the mechanism. Reactions of this type are classified as $S_{RN}1$ reactions (substitution radical nucleophilic unimolecular).

14   Autoxidation

14.1.   Introduction

Autoxidation can be defined as a radical-chain reaction between molecular oxygen and organic compounds at low or moderate temperatures. This reaction almost invariably results in the formation of hydroperoxides.

$$RH + O_2 \rightarrow ROOH$$

These hydroperoxides then frequently undergo further decomposition.

The autoxidation reaction may be beneficial or deleterious according to the particular circumstances under which it occurs. Thus, it is important in the drying of paints and oils as it leads to the formation of protective films, though these films ultimately break down as the result of further autoxidation. The oils used in paints contain unsaturated esters which give rise to hydroperoxides. Decomposition of these to alkoxy radicals by traces of added metal salts results in the formation of a protective polymer film by reaction of the alkoxy radical with further alkene.

Autoxidation is widely used in the petrochemical industry. Particularly important is the production of cumene hydroperoxide, acid-catalysed decomposition of which provides phenol and acetone:

$$PhCHMe_2 \xrightarrow{O_2} \underset{\underset{OOH}{|}}{PhCMe_2} \xrightarrow{H^+} PhOH + Me_2CO$$

The effects of autoxidation are highly undesirable in the rancidification of edible oils and the perishing of rubber. Petroleum and petroleum products are also susceptible to autoxidation giving rise to gummy products. In these cases it is necessary to find ways of controlling, or better inhibiting, autoxidation by suitable additives. These antioxidants are frequently phenols or amines.

14.2. Kinetic features of autoxidation

Alkanes, alkenes, alcohols, aldehydes and ethers are all susceptible to autoxidation by the same general process which involves the formation of a hydroperoxide:

$$RH + O_2 \xrightarrow{\text{initiator}} ROOH$$

The initiation, propagation and termination steps of this process are given by reactions (*1–5*):

$$\text{initiation} \begin{cases} \text{Initiator} \rightarrow 2\,\text{In} \cdot & (1) \\ \text{In} \cdot + RH \rightarrow R \cdot + InH & (2) \end{cases}$$

$$\text{propagation} \begin{cases} R \cdot + O_2 \rightarrow ROO \cdot & (3) \\ ROO \cdot + RH \rightarrow R \cdot + ROOH & (4) \end{cases}$$

$$\text{termination} \quad ROO \cdot + ROO \cdot \rightarrow \text{products} \qquad (5)$$

(i) *Initiation.* Initiation can be achieved by addition of typical radical initiators such as azonitriles and peroxides. The overall rate of autoxidation when a radical initiator is employed is proportional to the square root of the initiator concentration. In the absence of radical initiators, autoxidations are frequently subject to long induction periods. This induction period arises from the time needed for a build-up of hydroperoxide, thermolysis or photolysis of which produces radicals:

$$ROOH \rightarrow RO \cdot + HO \cdot$$

This hydroperoxide may arise as a result of interaction of singlet molecular oxygen with the organic substrate.

(ii) *Propagation.* The overall rate of autoxidation is independent of the oxygen pressure at moderate pressures of oxygen but is proportional to the concentration of the organic substrate RH. This is consistent with reaction (*3*) being fast and reaction (*4*) being rate limiting. One might expect reaction (*3*) to be fast since it is essentially a radical coupling process as molecular oxygen is a biradical. This step is probably diffusion controlled ($k \sim 10^9$ l mol^{-1} s^{-1}). The rate of reaction (*4*) is very susceptible to the chemical environment of the hydrogen being abstracted and the reaction does not occur at moderate temperatures ($<150°$C) unless the C—H bond undergoing abstraction is activated, e.g. the C—H bonds in ArCHR_2 and ArCHO. The reaction is faster, the more stable is R $\cdot$, the incipient radical. (The relative reactivities towards peroxy radicals of cyclohexane, toluene, cyclohexene and diphenylmethane are 0.014, 1.00, 18 and 30 respectively.) Polar and steric factors also influence the rate of this reaction.

The ROO—H bond strength is about 375 kJ mol^{-1}, which means that this bond is somewhat stronger than an allylic, benzylic or aldehydic C—H bond, but comparable to a tertiary C—H bond. This helps one to understand the observed selectivity of different sites in a molecule towards autoxidation. Reactions (*6–10*) illustrate this:

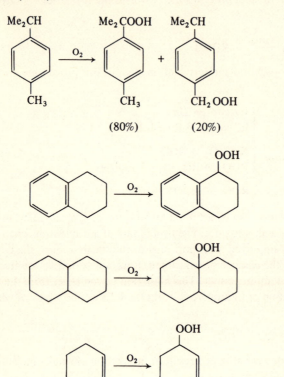

(80%)   (20%)

These examples illustrate that reaction occurs preferentially at allylic or benzylic positions, which give rise to a delocalized allylic or benzylic radical. Where no such position exists, as in decalin, the tertiary C—H bond is attacked; whilst in tetrahydrofuran the stabilizing effect of the oxygen on the resultant radical promotes attack at the α-C—H bond. Polar factors are also important here. The peroxy radical is an electrophilic radical, and hence in the transition state the α-carbon will bear a partial positive charge, which can be stabilized by the adjacent oxygen atom. Groups which increase the electrophilicity of the peroxy radical enhance its reactivity. Thus the radical (**1**) is much more reactive than the radical (**2**).

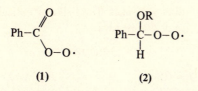

(1) (2)

Compounds which can give rise to more than one allylic radical such as methyl oleate, give mixtures of all the possible hydroperoxides:

$$CH_3(CH_2)_6CH_2CH=CHCH_2(CH_2)_6CO_2Me$$

$$\xrightarrow{O_2} \overset{\displaystyle OOH}{\underset{\displaystyle |}{CH_3(CH_2)_6CHCH=CH(CH_2)_7CO_2Me}}$$

$$+ \overset{\displaystyle OOH}{\underset{\displaystyle |}{CH_3(CH_2)_6CH=CHCH(CH_2)_7CO_2Me}}$$

$$+ \overset{\displaystyle OOH}{\underset{\displaystyle |}{CH_3(CH_2)_7CHCH=CH(CH_2)_6CO_2Me}}$$

$$+ \overset{\displaystyle OOH}{\underset{\displaystyle |}{CH_3(CH_2)_7CH=CHCH(CH_2)_6CO_2Me}}$$

(iii) *Termination.* The only terminal step which need be considered at moderate pressures of oxygen (above about 10 kPa, i.e. ∼0.1 atmosphere) is reaction *5*.

$$ROO\cdot + ROO\cdot \rightarrow Products \tag{5}$$

Reactions *11* and *12* involving alkyl radicals can be effectively ignored except at extremely low pressures of oxygen as the concentration of alkyl radicals is very low as a consequence of their very rapid reaction with oxygen.

$$R\cdot + ROO\cdot \rightarrow ROOR \tag{11}$$

$$R\cdot + R\cdot \rightarrow R-R \tag{12}$$

The rate constant for the bimolecular reaction of peroxy radicals is markedly dependent on the nature of the alkyl group of the peroxy radical. The chain termination rate constants for primary, secondary and tertiary alkylperoxy radicals are of the order 10^8, 10^6 and 10^3 l mol^{-1} s^{-1} respectively. The reactions involve the reversible formation of a dimer which is formulated as a tetroxide though tetroxides have never been isolated.

$$2ROO\cdot \rightleftharpoons ROOOOR \rightarrow Products$$

The tetroxide derived from t-alkylperoxy radicals undergoes O—O cleavage with the formation of oxygen and t-alkoxy radicals. The t-alkoxy radicals either dimerize within the solvent cage or diffuse away.

$$ROOOOR \longrightarrow [RO\cdot \quad O_2 \quad \cdot OR] \begin{cases} \nearrow ROOR + O_2 \\ \\ \searrow 2RO\cdot + O_2 \end{cases} \qquad (13)$$

Primary and secondary alkylperoxy radicals can behave similarly or alternatively the tetroxide may break down in a concerted fashion to give equimolar amounts of alcohol and a carbonyl compound (reaction *14*)

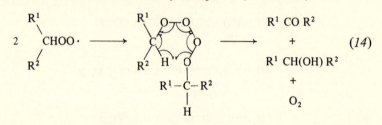

$$ (14) $$

The alkoxy radicals generated from the breakdown of tetroxides may initiate further chains (reaction *15*).

$$RO\cdot + RH \rightarrow ROH + R\cdot \qquad (15)$$

Alternatively the alkoxy radicals may disproportionate (reaction *16*) or cross-disproportionate with alkylperoxy radicals (reaction *17*): these processes are illustrated with benzyloxy radicals.

$$PhCH_2O\cdot + PhCH_2O\cdot \rightarrow PhCH_2OH + PhCHO \qquad (16)$$

$$PhCH_2O\cdot + PhCH_2OO\cdot \rightarrow PhCH_2OOH + PhCHO \qquad (17)$$

14.3. Catalysis of autoxidation by metal salts

Metal-catalysed decompositions of hydroperoxides can take place in two different ways according to the oxidation state of the metal:

$$ROOH + M^{n+} \rightarrow HOM^{n+} + RO\cdot$$

$$ROOH + M^{(n+1)+} \rightarrow ROO\cdot + M^{n+} + H^+$$

The effect of added metal salts is thus to initiate further radical chains, thereby increasing the overall rate of autoxidation. Use of larger quantities can, however sometimes inhibit autoxidation:

$$ROO\cdot + M^{n+} \rightarrow ROO^- + M^{(n+1)+}$$

Metal salts can also initiate the autoxidation process in the absence of hydroperoxide:

$$Co^{3+} + RH \rightarrow R\cdot + Co^{2+} + H^+$$

Thus, the rate of chain initiation in the oxidation of linoleic acid in benzene has been shown to be proportional to the concentration of added cobalt(III) stearate. This reaction is somewhat analogous to the oxidation of aromatic compounds by cobalt(III) acetate discussed earlier (p. 139).

14.4.　Antioxidants

The essential feature of an antioxidant is that it interrupts the autoxidation radical chain, thereby retarding the autoxidation process. It is thus necessary that compounds used as antioxidants shall have a readily abstractable hydrogen. This condition is satisfied by phenols and aromatic amines, both of which are extensively employed as antioxidants on account of their ability to act as efficient chain-transfer agents.

$$ROO\cdot + ArOH \rightarrow ROOH + ArO\cdot$$

$$ROO\cdot + ArNH_2 \rightarrow ROOH + Ar\overset{\cdot}{N}H$$

The kinetic scheme for autoxidation has to be modified in the presence of an antioxidant AH, by inclusion of reactions *18-21*.

$$ROO\cdot + AH \rightarrow ROOH + A\cdot \qquad (18)$$

$$A\cdot + ROO\cdot \rightarrow ROOA \qquad (19)$$

$$2A\cdot \rightarrow A{-}A \qquad (20)$$

$$A\cdot + RH \rightarrow R\cdot + AH \qquad (21)$$

The rate of reaction (*18*) is strongly dependent on steric factors (polar factors are also important). Thus 2,6-dialkylphenols react much less rapidly than simple phenols because of the steric protection of the reaction centre, and hence they are less efficient antioxidants (figure 14.1). For hindered phenols reactions (*19*) and (*20*) are much faster than reactions (*18*) and (*21*), whereas for simple phenols all four reactions are competitive. Thus, for hindered phenols the chain carrier is the peroxy radical but for non-hindered phenols the chain is perpetuated by both peroxy and phenoxy radicals. Consequently, non-hindered phenols can only retard but not inhibit autoxidation reactions.

A mixture of a hindered and unhindered phenol is, however, a much more efficient antioxidant than either alone as seen for a mixture of 2,6-di-t-butyl-4-methylphenol and *p*-methoxyphenol (see figure 14.1). The reason for this synergistic behaviour of the two phenols arises from the regeneration of *p*-methoxyphenol in reaction (*23*). This reaction occurs as shown and not in

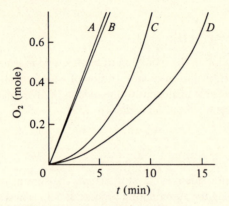

Figure 14.1. The oxygen absorption of a chlorobenzene solution of 148 mM 9,10-dihydroanthracene at 60°C containing 124 mM 2,2,3,3-tetraphenylbutane and various inhibitor concentrations. Curve *A*, no inhibitor; curve *B* 0.1 mM 2,6-di-t-butyl-4-methyl-phenol; curve *C*, 0.1 mM 4-methoxyphenol, curve *D* 0.1 mM 2,6-di-t-butyl-4-methyl-phenol and 0.1 mM 4-methoxyphenol. From L. R. Mahoney, *Angewandte Chemie* (International Edition), 1969, 8, 547; reproduced by permission of *Angewandte Chemie.*

the reverse direction, because of the relief of strain accompanying the formation of 2,6-disubstituted phenoxy radicals. In the absence of a hindered phenol reaction (22) would be reversible and consequently retardation would be less marked.

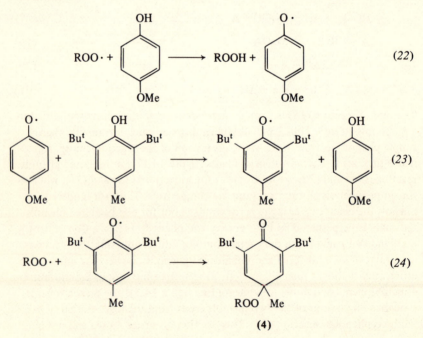

(4)

The peroxycyclohexadienone (4) is a typical cross-coupled product from a 2,4,6-trisubstituted phenoxy radical and a peroxy radical.

14.5. Autoxidation of aldehydes

Aldehydes are very susceptible to air oxidations by the radical-chain mechanism outlined in reactions (25–27).

$$PhCHO + R \cdot \longrightarrow Ph\overset{\cdot}{C}O + RH \qquad (25)$$

$$Ph\overset{\cdot}{C}O + O_2 \longrightarrow Ph{-}C\overset{\displaystyle O}{\underset{\displaystyle O{-}O \cdot}{}} \qquad (26)$$

$$PhC\overset{\displaystyle O}{\underset{\displaystyle O{-}O \cdot}{}} + PhCHO \longrightarrow Ph{-}C\overset{\displaystyle O}{\underset{\displaystyle O{-}O{-}H}{}} + Ph\overset{\cdot}{C}O \qquad (27)$$

The final product obtained from these reactions is generally the carboxylic acid rather than the peracid: this latter undergoes an acid-catalysed reaction with aldehyde (reaction *28*).

$$Ph{-}C\overset{\displaystyle O}{\underset{\displaystyle H}{}}$$

$$+ \qquad \xrightarrow{H^+} \quad Ph{-}\underset{\displaystyle O{-}OCOPh}{\overset{\displaystyle OH}{C}}{-}H \longrightarrow 2PhCOOH \qquad (28)$$

$$Ph{-}C\overset{\displaystyle O}{\underset{\displaystyle O{-}O{-}H}{}}$$

Some carbon monoxide is formed in autoxidations of aldehydes as a result of decarbonylation of the intermediate acyl radicals: the extent of this is greater with aldehydes containing secondary alkyl groups than with those possessing primary alkyl groups because of the greater stability of secondary alkyl radicals.

$$R\overset{\cdot}{C}O \rightarrow R \cdot + CO$$

Initiation of the autoxidation may be brought about photochemically, by radical initiators, or by metal ions:

$$RCHO \overset{h\nu}{\rightarrow} R\overset{\cdot}{C}O + H$$

$$RCHO + R' \cdot \rightarrow R\overset{\cdot}{C}O + R'H$$

$$RCHO + Co^{3+} \rightarrow R\overset{\cdot}{C}O + Co^{2+} + H^+$$

14.6. Autoxidation of alkenes

The propagation step in the autoxidations so far considered has been assumed to involve hydrogen abstractions from the substrate by the peroxy radical. In the case of alkenes this reaction (*29*) is accompanied by addition of a peroxy radical to the double bond (reaction *30*):

$$ROO\cdot + -\overset{|}{\underset{\underset{H}{|}}{C}}-\overset{|}{C}=C\diagup \longrightarrow ROOH + \overset{\diagdown}{C}-\overset{|}{C}=C\diagup \qquad (29)$$

$$ROO\cdot + \overset{\diagdown}{\underset{\diagup}{C}}=C\overset{\diagup}{\diagdown} \longrightarrow ROO-\overset{|}{C}-\overset{|}{\underset{\diagdown}{C}}\diagup \qquad (30)$$

The abstraction mechanism occurs almost exclusively for cyclopentene, cyclohexene, and open-chain alkenes with a tertiary allylic hydrogen, whilst the addition pathway predominates for alkenes with no reactive allylic hydrogens such as styrene and norbornene. The vast majority of alkenes react by both mechanisms to give mixtures of products.

The radical formed as a result of addition of a peroxy radical to a double bond reacts further by addition of oxygen or by fragmentation to give an epoxide and alkoxy radical:

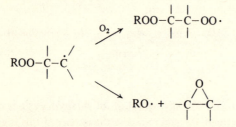

The greater the stability of the β-peroxyalkyl radical, the greater is its tendency to react with oxygen rather than to fragment.

The presence of oxygen during the polymerization of styrene has deleterious effects on the resultant polystyrene. The growing polymer radical may react with oxygen to give a polystyrylperoxy radical which reacts further to give, in addition to other products, a polyperoxide:

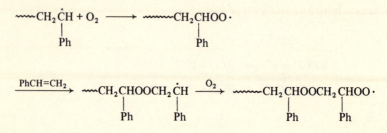

This polyperoxy radical can break down to give styrene oxide and a poly-alkoxy radical which 'unzips' to give benzaldehyde and formaldehyde:

$$\sim\sim\sim\left(OOCH_2\underset{\underset{Ph}{|}}{CH}\right)_n OOCH_2\overset{\cdot}{C}H \longrightarrow \sim\sim\sim\left(OOCH_2\underset{\underset{Ph}{|}}{CH}\right)_n O\cdot + \overset{O}{\overset{/\backslash}{CH_2-CHPh}}$$

$$\downarrow$$

$$n\text{PhCHO} + n\text{CH}_2\text{O}$$

15 Radical rearrangements

15.1. Introduction

In this chapter we shall be discussing reactions involving the rearrangement of radical intermediates and also molecular rearrangements which proceed via radical intermediates.

There are far fewer examples of rearrangements of radicals than of carbonium ions (see reactions *1* and *2*). This is due in part to the very much smaller energy difference between primary and tertiary radicals than between

$$Me_3CCH_2 \cdot \; \longrightarrow\!\!\!\times\!\!\!\longrightarrow \; Me_2\overset{\cdot}{C}CH_2Me \qquad\qquad (1)$$

$$Me_3CCH_2 + \; \longrightarrow \; Me_2\overset{+}{C}CH_2Me \qquad\qquad (2)$$

the corresponding carbonium ions. Another reason for the comparative paucity of radical rearrangements is the absence of a low-energy pathway for the 1,2-migration of alkyl groups.

The transition state (1) for a 1,2-alkyl migration in carbonium ion chemistry involves a π-complex between the double bond and the migrating alkyl group. Such a process is 'symmetry-allowed'. The extra electron in the transition state for the analogous radical rearrangement must perforce go into an anti-bonding orbital. In all examples involving the 1,2-migration of alkyl groups, the reactions have been shown to occur by an elimination addition mechanism and not by a concerted process. It is also relevant to note that 1,2-migrations of alkyl groups do not occur in carbanions.

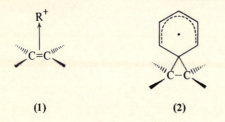

(1)   (2)

There are numerous examples of 1,2-aryl and vinyl migrations. These processes proceed via transition states such as (2), which are not subject to restrictions of orbital symmetry. Somewhat less clear is the precise mechanism for the facile 1,2-migrations of halogen atoms in β-haloalkyl radicals. The facile nature of these rearrangements is probably related to the availability of low energy *d* orbitals in the halogen: it is thus not necessary to place the unpaired electron in the transition state in a high energy antibonding orbital. 1,2-Migrations also occur by an elimination–addition pathway (reaction *3*) and by a mechanism proceeding through a five-membered transition state (reaction *4*).

$$MeCOCHMeCH_2\cdot \longrightarrow [Me\dot{C}O + MeCH{=}CH_2] \longrightarrow Me\dot{C}HCH_2COMe \quad (3)$$

$$\begin{array}{c} OCOMe \\ | \\ Me_2CCH_2\cdot \end{array} \longrightarrow \begin{array}{c} Me \\ | \\ C \\ O{\diagup}\;\;{\diagdown}O \\ \vdots\;\;\cdot\;\;\vdots \\ Me_2C{-}{-}CH_2 \end{array} \longrightarrow Me_2\dot{C}CH_2OCOMe \qquad (4)$$

15.2. Aryl migrations

The first rearrangement of a radical to be recognized was that of the primary neophyl radical ($PhCMe_2CH_2\cdot$) to the tertiary radical ($PhCH_2\dot{C}Me_2$). The decarbonylation of 3-methyl-3-phenylbutanal affords isobutylbenzene as well as t-butylbenzene. The extent of rearrangement is greatest under conditions where chain transfer with aldehyde is minimized, i.e. in dilute solution. Conversely the addition of a thiol to the reaction mixture results in the virtual suppression of all rearrangement (thiols are excellent chain-transfer agents).

$$PhCMe_2CH_2CHO \xrightarrow{Bu^tO\cdot} PhCMe_2CH_2\dot{C}O \xrightarrow{-CO} PhCMe_2CH_2\cdot \longrightarrow Me_2\dot{C}CH_2Ph$$

$$\left\downarrow \begin{array}{l} RCHO \\ or \\ R'SH \end{array} \qquad\qquad \left\downarrow \begin{array}{l} RCHO \\ or \\ R'SH \end{array}$$

$$PhCMe_3 \qquad\qquad Me_2CHCH_2Ph$$

These results indicate that discrete unrearranged and rearranged radicals are intermediates with bridged radicals being involved only as transition states or short-lived intermediates. This is further indicated from a study of the decarbonylation of the optically active aldehyde (3) to (4) which occurs with at least 98 per cent racemization (scheme 1). Had the bridged species (5) been a product-forming intermediate the product would have been optically active with inversion of configuration as occurs with the analogous phenonium ion.

The extent of rearrangement of β-arylalkyl radicals is dependent on the relative stabilization energies of the unrearranged and rearranged radicals.

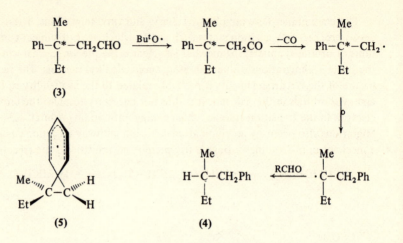

Scheme 1

Thus under comparable conditions the extent of rearrangement of $PhCH_2{}^{14}CH_2 \cdot$ is only 3 per cent while that of $Ph_3CCH_2 \cdot$ is complete.

$$PhCH_2{}^{14}CH_2 \cdot \rightarrow \dot{C}H_2{}^{14}CH_2Ph$$

$$Ph_3CCH_2 \cdot \rightarrow Ph_2\dot{C}CH_2Ph$$

The analogous 1,2-aryl shift to oxygen occurs in triarylmethoxy radicals.

$$Ph_3CO \cdot \rightarrow Ph_2\dot{C}OPh$$

δ-Arylalkyl radicals undergo 1,4-aryl migrations in addition to cyclization: the 1,4-aryl migration proceeds via an Ar_1-5 transition state (6) (scheme 2).

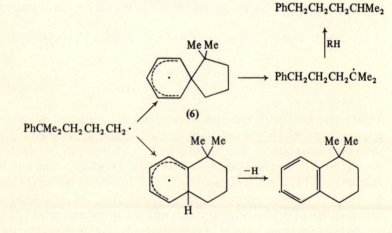

Scheme 2

Examples of 1,4-aryl migrations from carbon to oxygen and of 1,5-aryl shifts from oxygen to oxygen similarly occur by Ar_1-5 and Ar_1-6 transition states respectively. The driving force in each instance is the formation of a more stable radical.

15.3.   1,2-Vinyl migrations

Radical rearrangements involving the 1,2-migration of vinyl groups are comparatively common; the rearrangements proceeding via a cyclic species (see reaction 5).

$$CH_2=CHCHCH_2\cdot \longrightarrow \underset{\underset{Me}{\overset{\displaystyle CH}{|}}}{\overset{\displaystyle CH_2\cdot}{}} CH_2-CHMe \longrightarrow CH_2=CHCH_2\dot{C}HMe \quad (5)$$

The formation of both *cis*- and *trans*-1-deuteriobut-1-enes from the decarbonylation of the specifically labelled *cis*-deuteriated aldehyde (7) indicates that the rearrangement proceeds via the intermediate cyclopropylmethyl radical (8), and that this latter has sufficient lifetime to undergo rotation about the C—C bond before ring-opening, i.e. (8) is an intermediate and not merely a transition state.

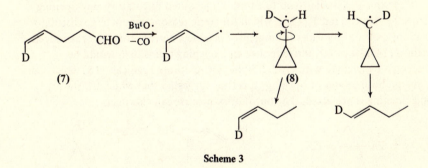

Scheme 3

Closely related to 1,2-migrations of vinyl groups are cyclopropylcarbinyl–homoallyl rearrangements. These may be exemplified by the formation of a 1 : 2 mixture of norbornene and nortricyclene from treatment of either (9) or (11) with tri-n-butyltin hydride at 25°C (see p. 164) (scheme 4). At lower temperatures the equilibration of the intermediate radicals (10) and (12) is incomplete, and the composition of the product mixture reflects the nature of the starting material.

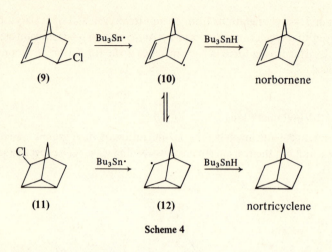

Scheme 4

15.4. Ring opening of cyclopropyl radicals

A lot of interest has been centred on the preferred mode of opening of cyclo-propyl radicals. Orbital symmetry requirements indicate that the preferred modes of ring opening of cyclopropyl cations and anions are disrotatory and conrotatory respectively. The molecular orbital picture is much less clear for the ring opening of cyclopropyl radicals: calculations indicate that ring-opening is formally forbidden in both modes but that a near-disrotatory mode is favoured for a distorted radical.

Experimental studies tend to support a favoured disrotatory ring-opening. Thus the bicyclic radical (**13**), which for steric reasons can undergo disrotatory but not conrotatory ring opening, gives products derived from the rearranged radical (**14**). However, the extent of ring opening is less than would be expected by analogy with *cis*-2,3-diphenylcyclopropyl radicals (**15**) which can undergo both modes of opening: it is thus suggested that while disrotatory ring opening is preferred, the conrotatory process can also occur.

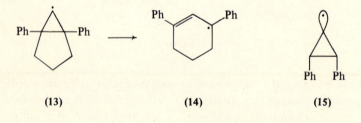

15.5. Migrations of hydrogen

The commonest type of hydrogen migration from carbon to carbon is the 1,5-shift (see reaction *6*) though examples of 1,3-, 1,4-, 1,6-, 1,10- and 1,11-

intramolecular hydrogen transfers are well authenticated:

$$(PhCH_2\,CH_2\,CH_2\,CH_2\,CH_2\,CO_2)_2 \overset{\Delta}{\rightarrow} 2PhCH_2\,CH_2\,CH_2\,CH_2\,CH_2 \cdot$$

$$CH_3\,CH_2\,CH_2\,CH_2\,CHPhCHPhCH_2\,CH_2\,CH_2\,CH_3 \leftarrow 2Ph\overset{\cdot}{C}HCH_2\,CH_2\,CH_2\,CH_3 \tag{6}$$

1,5-hydrogen migrations from carbon to oxygen occur even more readily as the new O—H bond is considerably stronger than the C—H bond which is broken (see reaction 7).

The Norrish type II process, so frequently observed in the photolysis of ketones, is another example of an intramolecular 1,5-hydrogen transfer (reaction 8).

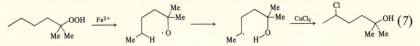

1,6-Hydrogen migration to oxygen also occurs though it is much less common than the analogous shift to carbon. Even when benzylic hydrogens are available for 1,6-transfer, 1,5-transfer predominates.

$$PhCH_2\,CH_2\,CH_2\,CH_2\,CMe_2\,OCl \xrightarrow{h\nu} PhCH_2\,CH_2\,CH_2\,CH_2\,CMe_2\,O \cdot$$

$$\left\{\begin{array}{c} PhCH_2\,CHClCH_2\,CH_2\,CMe_2\,OH \\ (73\ per\ cent) \\ \\ \\ PhCHClCH_2\,CH_2\,CH_2\,CMe_2\,OH \\ (7\ per\ cent) \end{array}\right\} \underset{(-RO\,\cdot)}{\overset{ROCl}{\longleftrightarrow}} \left\{\begin{array}{c} PhCH_2\,\overset{\cdot}{C}HCH_2\,CH_2\,CMe_2\,OH \\ \\ + \\ \\ Ph\overset{\cdot}{C}HCH_2\,CH_2\,CH_2\,CMe_2\,OH \end{array}\right\}$$

The ease with which 1,5-hydrogen transfer occurs is attributed to the fact that the transition state resembles the geometry of the chair form of a cyclohexane ring (see **16**): this allows favourable orbital overlap. When this is not possible, as for cyclohexyloxy radicals, no 1,5-hydrogen migration occurs. In this case the transition state would necessitate the cyclohexane adopting a boat conformation (**17**): the extra energy involved in this precludes hydrogen migration.

(16)　　　　　　　　　　**(17)**

For other types of hydrogen migration, eclipsing interactions in the transition state and/or increased distance between the radical centre and the H—C bond undergoing transfer militate against a facile reaction.

15.6.   The Barton reaction

1,5-Hydrogen migration in radicals forms the basis of an exceedingly valuable procedure for the functionalization of angular methyl groups in the steroid molecule to give compounds not readily accessible by other means. The essence of this reaction involves the generation of an alkoxy radical, by the photolysis of nitrite esters or less commonly of hypochlorites, in an environment such that 1,5-hydrogen migration can occur via a six-membered transition state. The resultant radical recombines with the nitric oxide produced in the photolysis of nitrites to give a nitroso-compound which subsequently tautomerizes to an oxime (scheme 5).

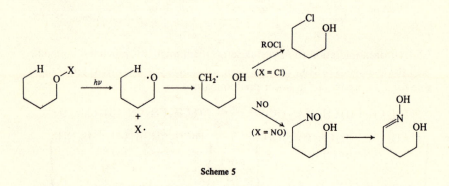

Scheme 5

The reaction has been successfully applied to the synthesis of C-18 and C-19 derivatives of steroids. The structure (18) indicates that C-18 derivatives should be formed by generation of the alkoxy radical at C-8, C-11, C-15 or C-20, and C-19 derivatives from a radical centre at C-2, C-4, C-6, C-8 or C-11. Schemes 6 and 7 illustrate examples of the Barton reaction as applied to the photolyses of nitrites and hypochlorites respectively.

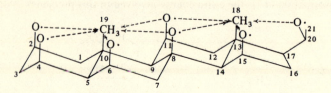

(18)

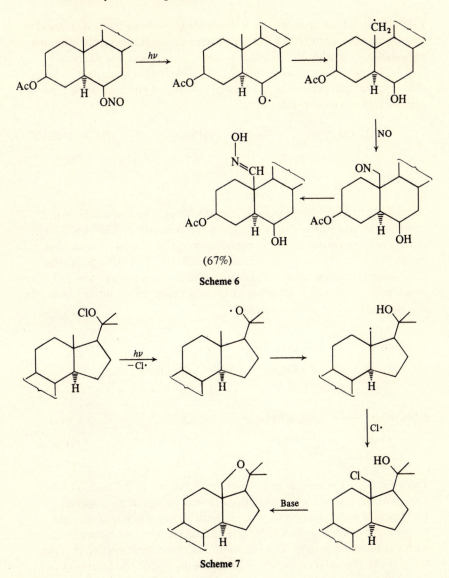

(67%)

Scheme 6

Scheme 7

15.7. Molecular rearrangements involving a radical-pair mechanism

Earlier in this chapter we considered rearrangements in which a radical intermediate rearranged to a more stable radical. In this section we shall turn our
attention to molecular rearrangements which proceed via a radical-pair mechanism: the substrate undergoes homolysis to a radical pair which recombine
within the solvent cage to give the rearranged product. In many instances

CIDNP studies have provided the evidence for the intermediacy of radicals, though the observation of CIDNP phenomena does not necessarily mean that radicals are involved in the principal product-forming reaction pathway.

The Stevens rearrangement of ylids (19) derived from quaternary ammonium salts (e.g. reaction (9)) is one of a class of rearrangements involving a 1,2-migration to an electron-rich centre.

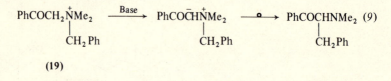

(19)

These rearrangements are intramolecular as indicated by 'cross-over' experiments. Rearrangements of optically active substrates occur with considerable retention of configuration of the migrating group. A concerted mechanism is unacceptable since it would be symmetry-forbidden. This leaves as possible mechanisms the ion-pair and radical-pair processes (reactions *10* and *11*), which must occur within the solvent cage to account for the intramolecularity and stereochemistry of the reaction.

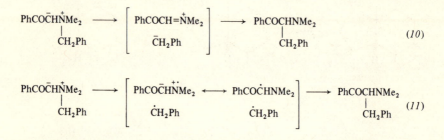

For a long time the ion-pair mechanism was favoured, as the reaction conditions, involving the use of strong base at elevated temperatures, seemed incompatible with a homolytic process. Relative migratory aptitudes and recently the observation of CIDNP phenomena are, however more consistent with a radical-pair process. This mechanism has been substantiated in some instances by the isolation of typical radical by-products, e.g. 5 per cent of bibenzyl was obtained in the rearrangement of (19). There are however, undoubtedly some examples of Stevens-type rearrangements in which ion pairs are involved and it may well be that both pathways on occasion occur concurrently.

The Wittig rearrangement involving the base-catalysed rearrangement of benzyl ethers is mechanistically very closely related to the Stevens rearrangement. CIDNP effects have been observed for several Wittig rearrangements supporting a radical mechanism.

$$\text{ArCHOR} \xrightarrow{\text{BuLi}} \underset{\underset{R'}{\mid}}{\text{Ar}\bar{\text{C}}\text{OR}} \longrightarrow \left[\underset{\underset{R'}{\mid}}{\text{Ar}\dot{\text{C}}-\text{O}^-} \overset{\text{R}\cdot}{} \right] \longrightarrow \underset{\underset{R'}{\mid}}{\overset{\overset{R}{\mid}}{\text{Ar}-\text{C}-\text{O}^-}}$$

A similar mechanism operates for the Meisenheimer rearrangement of amine oxides. In this example the yield of rearranged product is considerably reduced by carrying out the rearrangement in presence of oxygen or a thiol, both of which act as scavengers of the benzyl radical. The reaction carried out in

$$\underset{\underset{\text{Ph}}{\mid}}{\text{PhCH}_2\overset{+}{\text{N}}\text{Me}} \; \overset{O^-}{\overset{\mid}{}} \xrightarrow{\Delta} \left[\text{PhCH}_2\cdot \; \underset{\underset{\text{Ph}}{\mid}}{\overset{\overset{O\cdot}{\parallel}}{\text{N}-\text{Me}}} \right] \longrightarrow \text{PhCH}_2\text{ONMePh}$$

presence of a thiol gives toluene in a yield equivalent to the deficit in the yield of rearranged product. The second radical in the Wittig and Meisenheimer rearrangements is a ketyl radical anion and a nitroxide respectively, both of which are relatively long-lived species thus providing the driving force for the rearrangements.

16   Radicals in biological systems

16.1.   Introduction

In this chapter on radicals in biological systems we have selected four different areas in which the role of radicals is quite unambiguous. The examples chosen are ones in which there is a reasonable degree of certainty about the chemistry involved in the system and which illustrate different points. Inevitably the complexity of biological systems is such that a rather more qualitative approach has to be taken. This area of free-radical chemistry is undoubtedly one in which we can expect very significant developments in the near future.

16.2.   Biological oxidations

The general principles involved in enzyme action in biological oxidation will be discussed in this section. Attention will primarily be directed towards those processes involving single-electron transfer or its equivalent, hydrogen atom transfer. Transfer of two electrons or of hydride, which is equivalent to a two-electron transfer process, will only be mentioned in passing. Transfers of protons are neither oxidations nor reductions. Biological oxidations are effected not by enzymes, which are proteins, but by coenzymes which are complexed to the enzyme. The coenzyme functions as an electron carrier or as a carrier of hydrogen atoms or hydride ions.

   The energy requirements of animals and also of plant cells in the dark and of many microorganisms, are provided by the oxidation of organic substrates with molecular oxygen. There is, however, no simple chemical mechanism whereby this oxidation can be directly performed under mild conditions. To carry out the oxidation of an organic substrate likely to be encountered in the respiratory process it is necessary to break a C—H bond. Molecular oxygen is incapable of doing this under the conditions encountered in biological systems, as the reaction is endothermic. Nor is there any enzyme in mito-chondrial systems present in cells which can catalyse the direct reaction of the substrate with molecular oxygen.

$$SH_2 + \tfrac{1}{2}O_2 \rightarrow S + H_2O$$

This process can be carried out indirectly by the transfer of electrons from the substrate to a coenzyme.

$$SH_2 + \text{Coenz} \rightarrow S + \text{CoenzH}_2$$

The coenzyme generally involved in this process is nicotinamide-adenine dinucleotide (NAD^+) or its monophosphate ($NADP^+$). Oxidation of the reduced coenzyme with molecular oxygen does not occur directly but by a series of electron carriers which transport electrons to molecular oxygen from the substrate undergoing oxidation. Scheme 1 outlines a simplified sequence of the events occurring in the electron transport process involved in the oxidation of a substrate by molecular oxygen (where Fl represents a flavin nucleotide, FlH$_2$ the corresponding reduced flavin nucleotide, Q a ubiquinone and QH$_2$ the hydroquinone derived from the ubiquinone Q).

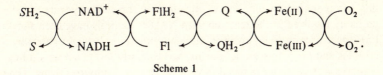

Scheme 1

Both one- and two-electron transfers (or their equivalents) are involved in the oxidation–reduction process. In some instances two concurrent one-electron processes may be occurring.

An electron transport chain is also involved in photosynthesis. The chief difference is that it then operates in the reverse direction.

Four types of oxidation–reduction enzymes participate in the mainstream of the electron-transport chain in the respiratory process. These are: (1) dehydrogenases which require either NAD^+ or $NADP^+$ as coenzyme; (2) flavin-linked dehydrogenases (Fl) which contain flavin adenine dinucleotide (FAD) or flavin mononucleotide (FMN); (3) iron–sulphur proteins and (4) cytochromes which possess an iron–porphyrin group. In addition the quinone, ubiquinone, is involved in the electron-transport sequence. The mode of action of the above will be discussed to illustrate some of the basic principles underlying the mechanism of enzyme action.

16.2.1. Oxidations with NAD$^+$

NAD^+ is normally a two-electron oxidant acting in most instances as a hydride acceptor. The amide group ($CONH_2$) in NAD^+ ensures that the coenzyme has the appropriate redox potential to bring about the oxidation of the organic substrate. The group R controls the binding of the coenzyme to the enzyme. $NADP^+$ functions similarly.

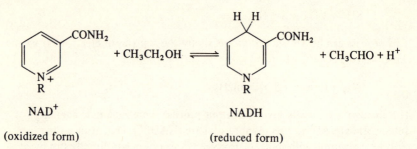

NAD⁺

(oxidized form)

NADH

(reduced form)

16.2.2. Oxidations with flavoproteins

The NADH formed in the oxidation of the organic substrate is reoxidized to NAD^+ by the flavoproteins (FAD or FMN).

$$NADH + Fl + H^+ \rightleftharpoons NAD^+ + FlH_2$$

The overall two-electron reduction of these flavoproteins can proceed in a single step or via two consecutive one-electron transfers: hydride transfer is probably involved in the reaction with NADH. Radicals have been shown by ESR spectroscopy to be involved in many oxidations brought about by flavoproteins and by simple model flavins, and similarly in the corresponding reductions effected by reduced flavoproteins or 1,5-dihydroflavins. The pathway clearly depends on the relative energies of the two alternative routes.

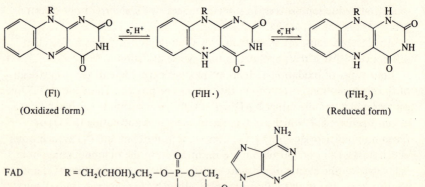

(Fl)

(Oxidized form)

(FlH·)

(FlH₂)

(Reduced form)

FAD $R = CH_2(CHOH)_3CH_2-O-\overset{\overset{\displaystyle O}{\|}}{\underset{\underset{\displaystyle OH}{|}}{P}}-O-CH_2$

FMN $R = CH_2(CHOH)_3CH_2-O-\overset{\overset{\displaystyle O}{\|}}{\underset{\underset{\displaystyle OH}{|}}{P}}-OH$

The radical route is of relatively low energy owing to the extensive delocalization of the unpaired electron in the flavin radical onto the carbocyclic ring and both nitrogens of the central ring (see (1)): there is little delocalization of spin onto the pyrimidine ring. This route appears to be preferred when the

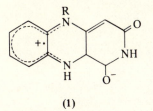

(1)

radical generated from the substrate is relatively stable as in reductions of simple carbonyl compounds: α-hydroxyalkyl radicals are relatively stable.

$$FlH_2 + R_2CO \rightarrow FlH\cdot + R_2\dot{C}OH$$

The detailed mechanism probably involves an electron transfer followed by exceedingly rapid proton transfer

$$FlH_2 + R_2CO \rightarrow FlH_2^{+\cdot} + R_2CO^{\overline{\cdot}} \rightarrow FlH\cdot + R_2\dot{C}OH$$

In contrast, hydride ion transfers are probably involved in the reduction of pyruvate to lactate. In this case the hydride-transfer route would be facilitated by the electron-withdrawing inductive influence of the adjacent ester group.

$$FlH_2 + MeCOCO_2R \longrightarrow FlH^+ + Me\overset{O^-}{\underset{|}{\dot{C}}}HCO_2R \longrightarrow Fl + MeCHOHCO_2R$$

NADH reductions normally proceed via the two-electron route since the unpaired spin in the radical derived from NADH cannot be delocalized to the same extent as in a flavin radical.

Some flavoproteins contain metals which are complexed to the flavin nucleotide. Xanthine oxidase contains molybdenum and iron. These metals are essential to the catalytic activity of the enzyme probably because facile single electron transfers can occur e.g. from iron(II) to iron(III).

16.2.3. Action of ubiquinone

Quinones are good one-electron oxidants forming the delocalized semiquinone radical anion. The radical anion can take up a proton and the resulting phenoxy radical then undergoes a further one-electron transfer and proton uptake to give a hydroquinone.

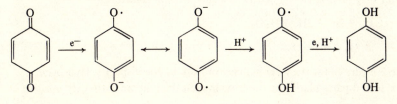

Ubiquinones (also known as coenzyme Q) act in the same way in the oxidation of reduced flavoproteins. The various ubiquinones differ only in the

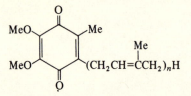

nature of the side chain (n may have several values). The function of the prenyl side chain is to facilitate binding to the membrane surface, while the two methoxy and the methyl groups ensure that the redox potential of the quinone is such that it can oxidize the dihydroflavins and at the same time the reduced quinone can itself be oxidized by iron–sulphur proteins.

16.2.4. *Action of the iron–sulphur proteins and cytochromes*

The iron–sulphur proteins and the cytochromes both act as electron carriers in biological oxidations by undergoing reversible iron(II)–iron(III) transitions.

$$Fe(III) + e^- \rightleftharpoons Fe(II)$$

Five different cytochromes are involved in the electron-transport chain in the mitochondria of higher animals. The redox potentials of these cytochromes become successively more positive at the oxygen terminus of the electron-transport chain. Only the final one (cytochrome a) is capable of undergoing an electron transfer with oxygen to give a superoxide radical anion.

$$O_2 + e^- \rightarrow O_2^{-\cdot}$$

16.2.5. *Reduction of oxygen*

Most of the molecular oxygen at the end of the electron-transport chain is converted into water. It is improbable that this occurs directly since this would involve a four-electron transfer.

$$O_2 + 4e^- + 4H^+ \rightarrow 2H_2O$$

More probable are two or one electron-transfer processes leading to hydrogen peroxide and the superoxide radical anion respectively: the two-electron transfer may involve two concurrent single-electron transfers.

$$O_2 + 2e^- + 2H^+ \rightarrow H_2O_2$$

$$O_2 + e^- \rightarrow O_2^{-\cdot}$$

There is evidence for the intermediacy of both of these species. Hydrogen peroxide and particularly the superoxide radical anion are extremely reactive and capable of causing irreversible damage to various biomolecules. That this does not occur is due to the superoxide dismutase enzyme and to the catalase and peroxidase enzymes which are present in aerobic cells. Superoxide dismutase converts the superoxide radical anion into hydrogen peroxide:

$$2\,O_2^{\cdot-} + 2H^+ \rightarrow O_2 + H_2O_2$$

The resultant hydrogen peroxide is then reduced by the associated catalase and peroxidase enzymes to water and oxygen. These enzymes are indispensable defences which make aerobic life possible.

A major difference in the mode of action of catalases and peroxidases is that the latter but not the former requires an external hydrogen donor such as

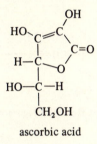

ascorbic acid

ascorbate or phenol. Both ascorbate and phenoxy radicals have been detected during oxidations using a peroxidase–hydrogen peroxide system. In the case of phenols, typical phenolic coupled products can be isolated (see p. 141).

Small amounts of molecular oxygen are also used to effect the hydroxylation of organic substrates. Hydroxylation is brought about by the haem-enzyme cytochrome P450. One oxygen atom is used in hydroxylation and the other is converted to water. These hydroxylation reactions bear some resemblance to hydroxylations with Fenton's reagent (p. 145) though from the pattern of hydroxylation of aromatic compounds it seems that free hydroxyl radicals are not the hydroxylating species.

16.3.   Autoxidation in biologically important compounds

In chapter 14 we discussed the general principles of autoxidation. Here we shall look at how autoxidation of biologically important compounds, particularly lipids, occurs.

The methylene-interrupted unsaturated carboxylic acid residues in the lipid portions of membranes are very susceptible to autoxidation on account of the allylic C−H bonds they possess (see p. 152).

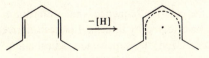

Autoxidation of the unsaturated components of lipids leads to lipid hydroperoxides, the breakdown of which gives rise to reactive radicals (see scheme 2). These radicals can then attack the associated protein or nucleic acid in the membrane thereby effecting its damage or breakdown.

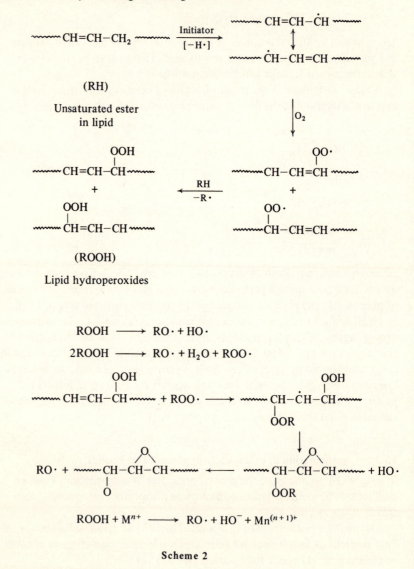

ROOH $\longrightarrow$ RO· + HO·

2ROOH $\longrightarrow$ RO· + H_2O + ROO·

ROOH + M^{n+} $\longrightarrow$ RO· + HO^- + $Mn^{(n+1)+}$

Scheme 2

The least-understood step in the reaction sequence is the initiation step. The hydrogen abstraction could be effected by hydroxyl radicals derived from hydrogen peroxide (cf. Fenton's reaction), by superoxide radical anions formed in the electron-transport chain (see p. 174), or in a photo-sensitized oxidation reaction by singlet oxygen or by triplet chlorophyll (or other pigment). Triplet chlorophyll behaves analogously to triplet benzophenone in readily abstracting hydrogens from substrates with weak C—H bonds (cf. p. 15).

$$Ph_2 CO \xrightarrow{h\nu} {}^3 Ph_2 CO*$$

$${}^3 Ph_2 CO* + RH \rightarrow Ph_2 \dot{C}OH + R\cdot$$

Lipid hydroperoxides are generated by the enzyme lipoxygenase. The reactions involve stereoselective removal of the hydrogen atom and isomerization of the double bond: the oxygen can attack the double bond from either side.

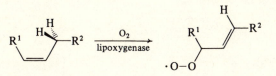

The bimolecular termination reaction of hydroperoxides or the reaction of peroxy radicals with hydroperoxide are only likely to be important processes with relatively high peroxide concentrations.

The unsaturated fatty acid components of lipids are largely protected from autoxidation by α-tocopherols (vitamin E). This reacts rapidly with alkylperoxy radicals to give a stable 2,4,6-trisubstituted phenoxy radical (see p. 50).

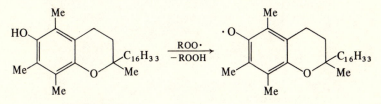

α-Tocopherol

There is some evidence that the α-tocopherol is regenerated by reaction of the phenoxy radical with the thiol tripeptide, glutathione (GSH), and the resultant glutathione disulphide reduced with NADPH.

$$ArO\cdot + GSH \rightarrow ArOH + GS\cdot$$

$$2GS\cdot \rightarrow GSSG$$

$$GSSG + NADPH + H^+ \xrightarrow[\text{glutathione} \atop \text{reductase}]{} 2GSH + NADPH^+$$

For the α-tocopherol to be at its most effective as an antioxidant it is necessary that it be bound to the membrane so as to be readily available to terminate the chain reaction: the C_{16} side chain facilitates binding.

Membrane damage due to lipid autoxidation has been associated with aging processes. It is possible that the extent of this damage can be decreased by vitamin E. In support of this, vitamin E has been shown to increase the life span of mice.

Lipid autoxidation in cells is not wholly a deleterious process: the prostaglandin hormones are obtained by autoxidation of the unsaturated C_{20} acid, arachidonic acid. This is essentially a controlled free-radical reaction involving cyclization of a peroxy radical to give, after further reaction with oxygen and hydrogen abstraction, the 15-hydroperoxy-endoperoxide PGG_2 (scheme 3).

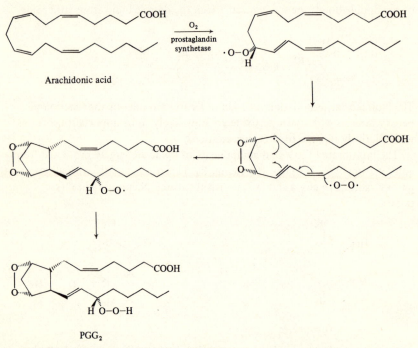

Scheme 3

Lipid autoxidation is responsible for objectionable odours and tastes in a variety of foodstuffs. Lipid hydroperoxides break down to give smaller relatively volatile substances, such as alcohols, aldehydes, ketones, acids, esters and lactones, which are responsible for the resulting flavours. Scheme 4 outlines how some of these compounds may arise from the 10-hydroperoxide of linoleates. The sequence involves decomposition of the hydroperoxide to an alkoxy radical which undergoes fragmentation in a direction such as to give the more stable allylic radical. This then reacts with oxygen to give a second hydroperoxide and thence a second alkoxy radical which gives *inter alia* pent-1-en-3-ol: this has been identified in oxidized butter.

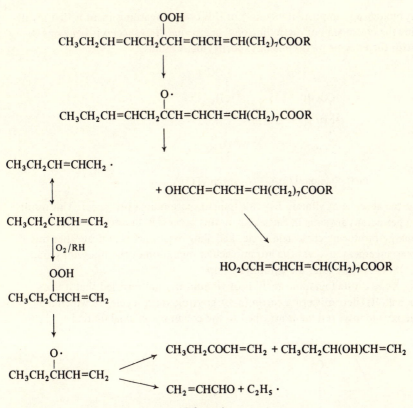

$$\text{OOH}$$
$$\text{CH}_3\text{CH}_2\text{CH=CHCH}_2\overset{|}{\text{C}}\text{CH=CHCH=CH(CH}_2)_7\text{COOR}$$

$$\downarrow$$

$$\text{O·}$$
$$\text{CH}_3\text{CH}_2\text{CH=CHCH}_2\overset{|}{\text{C}}\text{CH=CHCH=CH(CH}_2)_7\text{COOR}$$

$$\downarrow$$

$$\text{CH}_3\text{CH}_2\text{CH=CHCH}_2\text{·} \qquad\qquad + \text{OHCCH=CHCH=CH(CH}_2)_7\text{COOR}$$

$$\updownarrow$$

$$\text{CH}_3\text{CH}_2\overset{.}{\text{C}}\text{HCH=CH}_2$$

$$\downarrow \text{O}_2/\text{RH} \qquad\qquad\qquad \text{HO}_2\text{CCH=CHCH=CH(CH}_2)_7\text{COOR}$$

$$\text{OOH}$$
$$\text{CH}_3\text{CH}_2\overset{|}{\text{C}}\text{HCH=CH}_2$$

$$\downarrow$$

$$\text{O·}$$
$$\text{CH}_3\text{CH}_2\overset{|}{\text{C}}\text{HCH=CH}_2 \qquad \nearrow \text{CH}_3\text{CH}_2\text{COCH=CH}_2 + \text{CH}_3\text{CH}_2\text{CH(OH)CH=CH}_2$$
$$\searrow \text{CH}_2\text{=CHCHO} + \text{C}_2\text{H}_5\text{·}$$

Scheme 4

Off-flavours are also produced by γ-irradiation of foods, a procedure which has been used for preserving food. The irradiation induces C—C cleavage in the alkyl chains of lipids and other organic materials.

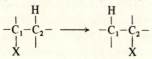

The resultant radicals are then converted to hydroperoxides and thence to alcohols, aldehydes etc. which contaminate the food.

16.4. Mechanism of action of coenzyme B$_{12}$

Coenzyme B$_{12}$ catalyzes a number of isomerizations. These all involve a 1,2-migration of a hydrogen atom and the reverse 2,1-migration of a group X, which may be OH, NH$_2$, alkyl or acyl.

$$-\overset{|}{\underset{X}{\text{C}}_1}-\overset{\overset{H}{|}}{\underset{|}{\text{C}}_2}- \longrightarrow -\overset{\overset{H}{|}}{\underset{|}{\text{C}}_1}-\overset{|}{\underset{X}{\text{C}}_2}-$$

A biologically important example of this type of rearrangement is that involving the conversion of methylmalonyl coenzyme A to succinyl coenzyme A with the enzyme methylmalonyl coenzyme A mutase.

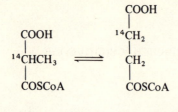

methylmalonyl CoA   succinyl CoA

In the absence of vitamin B_{12} this transformation does not occur. This results in pernicious anaemia in cattle, where this process is an essential adjunct to the energy-producing citric acid cycle. Labelling experiments indicate that this rearrangement involves the intramolecular migration of the thioester group COSCoA rather than the carboxyl group.

X-ray crystallographic analysis of vitamin B_{12} has revealed that it is a cobalt(III) derivative of a corrin (a tetrapyrrole macrocycle) to which a 5'-deoxyadenosyl residue is attached to the cobalt by an axial σ-bond.

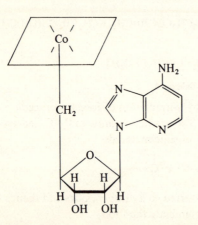

The conversion of propane-1,2-diol to propanal with the coenzyme B_{12}–diol dehydrase system is perhaps the best-understood example of a rearrangement catalyzed by an adenosylcobalamin coenzyme.

$$CH_3 CH(OH)CH_2 OH \rightarrow CH_3 CH_2 CH(OH)_2 \xrightarrow{-H_2O} CH_3 CH_2 CHO$$

There is considerable evidence that this and other similar enzyme-catalysed reactions using adenosyl cobalamin involve homolytic cleavage of the cobalt–carbon bond in the coenzyme to give a 5'-deoxyadenosyl radical and a cobalt(II) species: this latter has been detected by its characteristic ESR signal.

The adenosyl radical then abstracts a hydrogen atom from the substrate to give 5′-deoxyadenosine and the substrate radical. After migration of the group X (the OH group in the case of propane-1,2-diol), the rearranged radical retrieves a hydrogen from the 5′-deoxyadenosine to give the rearranged substrate and the adenosyl radical. This recombines with the cobalt(II) species to regenerate the coenzyme. Labelling studies have confirmed that the enzyme-adenosylcobalamin complex serves as an intermediate hydrogen carrier, first accepting a hydrogen from the substrate and then in a subsequent step transferring it back to the product. The hydrogen atoms of the substrate and the product do not exchange with the solvent: steric effects preventing solvent

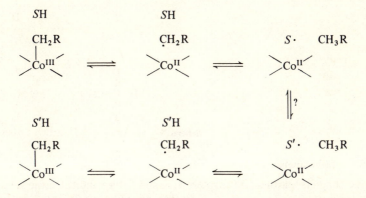

access to the reaction site. The mechanism of the step involving migration of the group X is not well understood and does not necessarily involve a simple 1,2-rearrangement of the substrate radical $S \cdot$ to the rearranged radical $S' \cdot$.

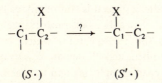

An alternative is that the substrate radical combines with the divalent cobalt(II) species and the intermediate organocobalt complex undergoes rearrangement. The rearranged substrate radical $S' \cdot$ does seem to be a distinct intermediate in the ethanolamine ammonia-lyase deamination of chirally labelled [2-^{2}H-2-^{3}H]ethanolamine to acetaldehyde (scheme 5). In this reaction the chirality of the labelled carbon atom is lost, as would be expected if protonated 1-amino-1-hydroxy-2-ethyl radicals (2) (the rearranged radicals) are intermediates in the reaction with sufficient lifetime to undergo rotation about the C_1-C_2 bond.

There are a number of other biological processes catalysed by other vitamin B₁₂ coenzymes but these appear to proceed by heterolytic mechanisms.

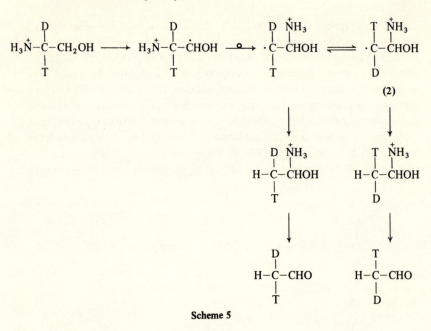

Scheme 5

16.5. Spin labels

Earlier chapters, particularly chapters 3, 5, and 6, have indicated the value of ESR spectroscopy as a means of establishing both the nature and also the conformations of organic radicals. Most biological systems are not paramagnetic and therefore do not give an ESR signal. Information about the conformations of the large complex biological molecules present in such systems can, however, be gained by attaching a radical, or as it is commonly called a spin label, to a natural component present. The spin labels most frequently used are nitroxides, related to (3) and (4), the ESR spectra of which consist of simple 1 : 1 : 1 triplets (see p. 22). The nitroxides chosen are always ditertiary nitroxides to avoid any further hyperfine splitting and to ensure the

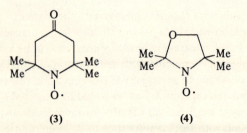

stability of the radical (p. 50). It is essential that the spin probe does not perturb the system to the extent that the structural data are irrelevant to the system under study.

The hyperfine splitting constants and the *g*-factors of radicals depend in the general case on their orientation with respect to the applied magnetic field. These anisotropic effects are averaged for small radicals in solution, where rapid tumbling occurs, to give spectra which are essentially isotropic. Such averaging is not the case for radicals embedded in single crystals where the ESR parameters are different for different orientations of the radical to the direction of the magnetic field (see figure 16.1) nor for radicals in viscous

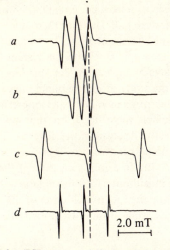

Figure 16.1. ESR spectra of di-t-butyl nitroxide in a single host crystal. The nitroxide molecule is orientated so that the N–O bond and the *p*-orbital are parallel to the *x* and *z* axes respectively. The crystal is rotated to align the applied magnetic field with the *x*, *y* and *z* axes of the nitroxide to give spectra (*a*), (*b*) and (*c*). The spectrum (*d*) is obtained from the nitroxide as a solution in di-t-butyl ketone. The vertical dotted line is the *g* value = 2.00036. From O. H. Griffith and A. S. Waggoner, *Accounts of Chemical Research*, 1969, **2**, 17; reproduced by permission of *Accounts of Chemical Research*.

solutions or in other media where motional averaging is frequently incomplete. It is this inherent anisotropy of the spin label which is utilized in the study of the orientation and movement of spin-labelled biological molecules particularly in membrances and phospholipids.

Spin labels can be attached to proteins by reaction of thiol or amino groups with the iodoacetamide (**5**) and maleimide nitroxides (**6**) respectively.

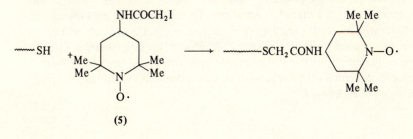

(5)

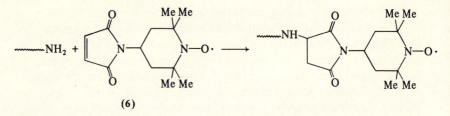

(6)

In this way nitrogenase enzymes have been spin labelled. Examination of the variation of the ESR spectra with temperature indicates that there is a definite change in the conformation of the enzyme at 19°C.

An alternative approach is to intercalate a spin label into the system under study, i.e. it is interposed between the large biomolecules. The spin label, 12-doxylstearic acid, is used in this way in the study of phospholipid bilayers and membranes. When the label is intercalated into a membrane, its long-axis label will be orientated along the length of the lipid molecule. It can thus move solely about its long axis, i.e. anisotropically. Analysis of the differing ESR spectra taken both along the lipid axis and perpendicular to it enables the movement of the spin label to be studied: this in turn is a measure of the motion of the lipid. The rate of flipping in and out of the membrane by the lipid label has been estimated by measuring the rate of reduction of the nitroxide by ascorbic acid: this is only accessible at the solvent membrane interface.

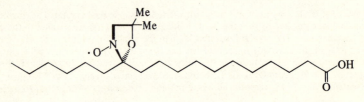

12-doxylstearic acid

In most spin-label experiments the spin labels are essentially isolated from each other, either because they are attached to well-separated positions on a large macromolecule, or, as in the case of membrane and lipid systems, they are present at very low concentration (about 1 mole per cent). When spin labels come into close proximity, interaction between the labels leads to line broadening and/or a change in line shape. The magnitude of such interactions falls off with increasing separation of the spin labels. Use is made of this effect in measuring the rate of lateral diffusion of lipid molecules in membranes by monitoring the change in spin–spin interaction from some local high concentration of the spin label, as this commences to diffuse uniformly through the membrane.

17   Radical displacement reactions

17.1.   Nature of the reaction

We are concerned in this chapter with reactions in which a radical attacks a multivalent centre M with displacement of a group which might be a single atom or a polyatomic radical:

$$R \cdot + MX_n \rightarrow RMX_{n-1} + X \cdot \qquad S_H 2$$

This type of process is formally similar to the well-known nucleophilic substitution reaction, and is labelled $S_H 2$ by analogy.

$$Nu^- + R-X \rightarrow NuR + X^- \qquad S_N 2$$

We exclude from our study the displacement reactions of radicals at hydrogen or halogen i.e. abstraction reactions:

$$R \cdot + H-X \rightarrow RH + X \cdot$$

since these are covered elsewhere. We shall concentrate on homolytic reactions where M is a multivalent centre.

Radical displacements can be expected to take place at atoms having vacant p or d orbitals available for coordination with the incoming radical. It is found that the reaction is important at elements of Groups II and III of the periodic table, for example at boron:

$$Me \cdot + BEt_3 \rightarrow MeBEt_2 + Et \cdot$$

Similarly, displacements readily occur at elements of the second and third rows of the periodic table which have energetically accessible d orbitals, for example at phosphorus:

$$Bu^t O \cdot + Me_3 P \rightarrow Bu^t OPMe_2 + Me \cdot$$

and at sulphur:

$$Ph \cdot + EtSSEt \rightarrow PhSEt + EtS \cdot$$

A weak bond between the leaving group and the multivalent centre, such as the S—S bond in the above disulphide, also facilitates the process.

Displacement reactions at saturated (sp^3-hybridized) carbon atoms are very rare under the conditions of temperature and pressure normally available to the organic chemist. Almost the only known examples of displacement reactions at carbon are the halogenations of strained rings, particularly cyclopropane and its derivatives to give 1,3-dihalopropanes.

$$\triangle \xrightarrow{\text{X} \cdot} {}_{\text{X}} \cdots \triangle \longrightarrow XCH_2CH_2CH_2 \cdot \xrightarrow{\text{X}_2} XCH_2CH_2CH_2X$$

One difficulty in the identification of S_H2 reactions is that the same overall result may be obtained by homolysis of the substrate followed by combination of one of the fragments with the attacking radical. For example, in the reaction of phenyl radicals with ethyl disulphide the product sulphide might be formed as follows:

$$EtSSEt \rightarrow 2EtS \cdot$$
$$EtS \cdot + Ph \cdot \rightarrow EtSPh$$

rather than in the direct substitution by phenyl on sulphur. Normally, product analysis alone is not sufficient to establish the mechanism, and additional kinetic or spectroscopic evidence is required.

17.2.   Displacement at group II and group III elements

The boron atom has a vacant p-orbital in the valence shell, so that an addition complex between an organoborane and an incoming radical seems feasible. Experimentally it is found that a variety of alkyl, alkoxy and acetyl radicals can displace alkyl radicals from alkylboranes e.g.:

$$MeO \cdot + Pr^i_3B \rightarrow Pr^i \cdot + Pr^i_2BOMe$$

The ESR spectra of the displaced alkyl radicals can be observed during the photolysis of a mixture of the peroxide and the alkylborane. The reactivities of the alkylboranes in this process decrease in the order $Bu^n_3B > Bu^i_3B > Bu^t_3B$, which suggests that the reaction is controlled by the steric congestion around the boron atom.

Dialkylmagnesium compounds and Grignard reagents react with t-butoxy radicals to give both alkyl and 1-magnesio-alkyl radicals:

$$R_2CHMgX + Bu^tO \cdot \left\langle \begin{array}{l} R_2CH \cdot + Bu^tOMgX \\ R_2\dot{C}MgX + Bu^tOH \end{array} \right.$$

The ESR spectrum of the solution shows a superposition of the spectra of the two radicals. The alkyl radical is formed by a homolytic displacement at magnesium.

17.3. Displacement at group IV elements

Radical substitution reactions are reasonably well-established for all the elements of group IV except carbon, for which homolytic displacement does not readily occur, except in the halogenation of cyclopropane as explained above. Alkyl radicals, such as trifluoromethyl, substitute silanes,

$$CF_3 \cdot + Me_4Si \rightarrow Me_3SiCF_3 + Me \cdot$$

and halogen atoms are known to displace silyl radicals from disilanes.

$$I \cdot + Me_3SiSiMe_3 \rightarrow Me_3SiI + Me_3Si \cdot$$

Displacement occurs at tin in a variety of organotin compounds under the mildest conditions. The weak $Sn-Sn$ bond in alkyl ditins is readily cleaved by alkyl, alkoxy and other radicals:

$$Bu^tO \cdot + Bu^n_3SnSnBu^n_3 \rightarrow Bu^tOSnBu^n_3 + Bu^n_3Sn \cdot$$

An intriguing variation on this reaction involves addition of an alkyl radical to a trialkylallyltin compound. The radical adds to the terminus of the allyl double bond and the allyl group is then displaced from the tin:

$$R \cdot + Bu^n_3SnCH_2CH=CH_2 \rightarrow Bu^n_3SnCH_2\dot{C}HCH_2R \rightarrow RCH_2CH=CH_2$$
$$+ Bu^n_3Sn \cdot$$

17.4. Displacement at group V elements

Displacement at nitrogen is relatively unfavourable because it has no vacant p or low-lying d-orbitals: only a few examples are known. When a radical attacks at phosphorus or arsenic these elements expand their valency from III to IV and the radical adds to give a phosphoranyl or arsanyl radical. Alkoxy radicals add to phosphines to give phosphoranyl radicals which subsequently undergo either α-scission, in which an alkyl radical is displaced from the phosphorus, or β-scission giving a phosphine oxide. In the case of α-scission the overall reaction is a homolytic displacement, and the intermediate phosphoranyl radicals are sufficiently long-lived to be detected by ESR. The alternative β-scission route is not a homolytic substitution.

$$RO \cdot + PX_3 \longrightarrow RO\dot{P}X_3 \quad \overset{\alpha\text{-scission}}{\underset{\beta\text{-scission}}{\diagup \diagdown}} \quad \begin{matrix} ROPX_2 + X \cdot \\ O = PX_3 + R \cdot \end{matrix}$$

The ESR spectra of the phosphoranyl radicals and of the displaced alkyl radicals can often be observed directly, or as adducts with a spin trap such as t-nitrosobutane (see p. 50).

A wide variety of phosphoranyl radicals have been prepared by addition of alkoxy radicals, alkyl radicals, halogen atoms, etc. to alkylphosphines R_3P, alkyl phosphites $(RO)_3P$, amino phosphines $(R_2N)_3P$ and phosphorus tri-halides. The phosphoranyl radicals generally have trigonal–bypyramidal struc-tures with the unpaired electron occupying an equatorial position (see p. 44). The most electrophilic ligands tend to occupy the apical (axial) positions.

The relative importance of α-scission and β-scission appears to depend mainly on the relative strengths of the bonds being broken, i.e. P—X or R—O. The β-scission route is favoured at higher temperatures, and in more polar solvents, and is the exclusive mode of decomposition of trialkylphosphites.

$$Bu^tO\cdot + P(OR)_3 \rightarrow Bu^t\cdot + O{=}P(OR)_3$$

The rate increases in the order: R = primary < secondary < tertiary. In tri-phenyl phosphite the α-scission route also occurs. As might be expected from the greater length of the bond, alkyl groups depart more readily from apical positions in the α-scission process.

17.5.   Displacement at group VI elements

Many displacement reactions at oxygen in peroxidic compounds have been identified, but not in other oxygen-containing compounds, probably because the O—C and O—H bonds are strong in contrast to the weak O—O bond. Peroxides decompose by a straightforward unimolecular homolysis:

$$ROOR \rightarrow 2RO\cdot$$

but this is frequently accompanied by a higher order 'induced' decomposition. This 'induced' reaction occurs by homolytic substitution at the peroxidic oxygen.

$$X\cdot + ROOR \rightarrow XOR + RO\cdot$$

The occurrence of the induced decomposition is critically dependent on factors such as the solvent (from which the attacking radical is derived), and substituents in the peroxide. Delocalized radicals attack peroxidic oxygen in this way, as for example in the induced decomposition of dibenzoyl peroxide in benzene which is effected by phenylcyclohexadienyl radicals (see also chapter 2).

$$(PhCO_2)_2 \longrightarrow PhCO_2\cdot \longrightarrow Ph\cdot + CO_2$$

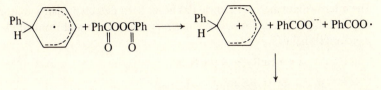

$$PhPh + PhCO_2H + PhCOO \cdot$$

The electron-rich peroxidic oxygen in peroxides is susceptible to S_H2 attack by nucleophilic radicals. For example, in the decomposition of dibenzoyl peroxide in primary or secondary alcohols, α-hydroxyalkyl radicals are the attacking species:

$$CH_3\dot{C}HOH + (PhCO_2)_2 \rightarrow PhCO_2CH(OH)CH_3 + PhCO_2 \cdot$$

$$PhCO_2CH(OH)CH_3 \rightarrow PhCO_2H + CH_3CHO$$

Dialkyl peroxides also undergo induced decomposition in the same way. Di-t-butyl peroxide undergoes a homolytic displacement by α-hydroxyalkyl or α-aminoalkyl radicals in alcohols or amines as solvents.

$$R_2\dot{C}OH + Bu^tOOBu^t \rightarrow Bu^tOH + R_2CO + Bu^tO \cdot$$

Radical displacement at sulphur is, perhaps, the best-established, and best-known example of homolytic substitution. Displacement at the weak S—S bond in disulphides has been observed for attack by alkyl, aryl, vinyl, thiyl, and tin radicals:

$$Ph \cdot + CH_3SSCH_3 \rightarrow PhSCH_3 + CH_3S \cdot$$

$$Bu^nS \cdot + MeC{\equiv}CH \rightarrow Bu^nSCH{=}\dot{C}Me$$

$$Bu^nSCH{=}\dot{C}Me + Bu^nSSBu^n \rightarrow Bu^nSCH{=}CHMeSBu^n + Bu^nS \cdot$$

$$Bu^n_3Sn \cdot + PhCH_2SSCH_2Ph \rightarrow Bu^n_3SnSCH_2Ph + PhCH_2S \cdot$$

Displacement also takes place in sulphides, but only when the bond between sulphur and the departing radical is relatively weak.

Quantitative estimates of the rates of reaction of dialkyl disulphides towards phenyl, polystyryl and tri-n-butyltin radicals have shown that the rate of displacement increases as the size of the alkyl substituents on sulphur decreases. The rate is a minimum at di-t-butyl disulphide for all three radicals. The indications are that steric effects are most important in controlling the reactivity.

17.6. Displacement at mercury and transition metal atoms

Radical displacement reactions at mercury are well established and may occur by addition of the attacking radical to the mercury atom, which expands its valence to III.

$$R \cdot + HgX_2 \rightarrow R\dot{H}gX_2 \rightarrow RHgX + X \cdot$$

The intermediate mercury radical has never been detected. Halogen atoms displace alkyl radicals from dialkylmercury compounds and from alkyl-mercury(II) halides:

$$I\cdot + HgR_2 \rightarrow R\cdot + RHgI$$

$$Br\cdot + Bu^sHgBr \rightarrow Bu^s\cdot + HgBr_2$$

The homolytic substitution competes with ionic reactions and non-polar solvents must be used. Alkyl and aryl radicals also displace alkyl radicals from organomercury compounds. For example, *cis*- and *trans*-4-methylcyclohexyl-mercury(II) bromides, when treated with dibenzoyl peroxide, gave an identical mixture of *cis*- plus *trans*-4-methylcyclohexylmercury(II) bromides, showing that the reactions proceed via a common radical intermediate.

$$(PhCO_2)_2 \rightarrow 2PhCO_2 \rightarrow 2Ph\cdot + 2CO_2$$

$$Ph\cdot + cis\text{-}RHgBr \rightarrow PhHgBr + R\cdot$$

$$R\cdot + cis\text{-}RHgBr \rightarrow cis\text{-} + trans\text{-}RHgBr + R\cdot$$

Homolytic displacement reactions at platinum have recently been discovered, ESR spectra have been recorded of the alkyl radicals (and their spin-adducts with Bu^tNO) displaced from platinum(II) complexes, PtR_2X_2 by alkoxy and thiyl radicals:

$$cis\text{-}PtR_2X_2 + Bu^tO\cdot \rightarrow PtR(OBu^t)X_2 + R\cdot$$

where R represents an alkyl group and X a phosphine ligand.

Thiyl radicals react with methylgold(I) and methylplatinum(II) complexes by a process involving a homolytic substitution at the metal centre:

$$MeAu + PhS\cdot \longrightarrow [Me\overset{\cdot}{A}uSPh] \begin{cases} \nearrow AuSPh + Me\cdot \\ \searrow PhS\cdot + Me(H)AuSPh \end{cases}$$

(branch labelled PhSH)

17.7. Stereochemistry of homolytic displacement reactions

It is well established that bimolecular nucleophilic substitution at saturated carbon atoms proceeds with inversion of configuration and, by analogy, similar stereochemistry for the homolytic substitution reaction might be expected. The attacking radical could approach from the backside and the configuration would be inverted.

Unfortunately, it has not been possible to test this hypothesis directly because no examples of homolytic displacement at chiral carbon atoms are known.

In an indirect approach to the stereochemical question the rates of homolytic substitutions at sulphides were compared with those of analogous nucleophilic substitutions known to proceed with inversion of configuration.

$$Y \cdot + RSSR \rightarrow YSR + \cdot SR$$

$$Y^- + RSX \rightarrow YSR + X^-$$

$$Y^- + RCH_2 X \rightarrow YCH_2 R + X^-$$

As the alkyl substituents were varied, linear correlations of the relative rates were obtained. This appeared to support the idea that the homolytic displacement at sulphur proceeded with inversion. Subsequently, it was shown that a large number of reactions, some of which were known to proceed without inversion, could also be correlated with the nucleophilic substitution rates. The range of alkyl substituents was simply not great enough for the correlations to have any mechanistic or stereochemical significance.

It seems reasonable, nevertheless, that homolytic substitution should proceed with inversion. The lowest energy route would involve a trigonal–bipyramidal intermediate in which the electron repulsion between the breaking and forming bonds and the equatorial electron pairs is minimized. With disulphides, for example, the intermediate would have structure (1):

$$Y \cdot + RSSR \longrightarrow Y \cdots\overset{\overset{\textstyle R}{|}}{S}\cdots SR \longrightarrow YSR + \cdot SR$$

(1)

Phosphoranyl radicals, which are the relatively stable intermediates formed with organophosphorus compounds, are known to have trigonal–bipyramidal structures. This arrangement of atoms is certainly consistent with a mechanism involving inversion of configuration.

Evidence that the homolytic substitution at a cyclopropane carbon involves inversion of configuration has been provided by studies of bromination of substituted cyclopropanes. For example, halogenation of *trans*-2,3-dideuterio-1,1-dichlorocyclopropane (2) with either bromine or chlorine gave more than 96 per cent of the *erythro*-product (3), and similar yields of the *threo*-product were obtained from the *cis*-dideuterio-isomer. The mechanism must therefore involve inversion as shown.

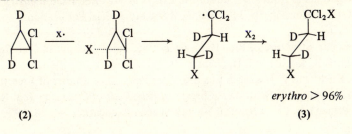

erythro > 96%

(2)   **(3)**

Suggestions for further reading

General references

D. C. Nonhebel and J. C. Walton. *Free-Radical Chemistry*, Cambridge University Press, 1974.

J. K. Kochi (ed.). *Free Radicals*, vols. I and II. Wiley, New York, 1973.

W. A. Waters (ed.). *Free-Radical Reactions.* (International Review of Science, vols. I and II.)

Annual Reviews on Free Radical Reactions. In *Organic Reactions Mechanisms*, Wiley, London.

Electron Spin Resonance, Chemical Society Special Publications, London.

W. A. Pryor. *Free Radicals*, McGraw-Hill, New York, 1966.

E. S. Huyser. *Free Radical Chain Reactions*, Interscience, New York, 1970.

Chapter 3

B. J. Tabner. Electron Spin Resonance Spectroscopy, in *Spectroscopy* (eds. B. P. Straughan and S. Walker), chapter 4. Chapman and Hall, London, 1976.

L. Kevan. Free-Radical Study by Electron Paramagnetic Resonance. In *Methods in Free-Radical Chemistry* (ed. E. S. Huyser), vol. I, chapter 1, 1969.

M. C. R. Symons. The Identification of Organic Free Radicals by Electron Spin Resonance. *Advances in Physical-Organic Chemistry*, 1963, 1, 284.

R. O. C. Norman. Electron Spin Resonance Studies of Free Radicals and Their Reactions in Aqueous Solution. In *Essays in Free-Radical Chemistry* (ed. R. O. C. Norman), p. 117. Chemical Society Special Publication No. 24, 1970.

H. Fischer. Structure of Free Radicals by ESR Spectroscopy. In *Free Radicals* (ed. J. K. Kochi), vol. II, chapter 19. Wiley, New York, 1973.

J. K. Kochi and P. J. Krusic. Electron Spin Resonance Studies of Free Radicals in Non-aqueous Solution. In *Essays in Free-Radical Chemistry* (ed. R. O. C. Norman), p. 147. Chemical Society Special Publication No. 24, 1970.

M. J. Perkins. The Trapping of Free Radicals by Diamagnetic Scavengers. In *Essays in Free-Radical Chemistry* (ed. R. O. C. Norman), p. 97. Chemical Society Special Publication No. 24, 1970.

C. Lagercrantz. Spin Trapping of Some Short-Lived Radicals by the Nitroxide Method. *Journal of Physical Chemistry, 1971, 75,* 3466.

Chapter 4

H. R. Ward. Chemically Induced Dynamic Nuclear Polarization (CIDNP). I. The Phenomenon, Examples, and Applications, *Accounts of Chemical Research, 1972,* **5,** 18.

R. G. Lawler. Chemically Induced Dynamic Nuclear Polarization (CIDNP). II. The Radical-Pair Model, *Accounts of Chemical Research*, 1972, **5,** 25.

R. Kaptein. Chemically Induced Dynamic Nuclear Polarization: Theory and Applications in Mechanistic Chemistry, *Advances in Free-Radical Chemistry*, 1975, **5,** 319.

Chapter 5

L. Kaplan. The Structure and Stereochemistry of Free Radicals. In *Free Radicals* (ed. J. K. Kochi), vol. II, chapter 18. Wiley, New York, 1973.

J. K. Kochi. Configuration and Conformation of Transient Alkyl Radicals in Solution by Electron Spin Resonance. *Advances in Free-Radical Chemistry*, 1975, **5,** 189.

O. Simamura. Stereochemistry of Cyclohexyl and Vinylic Radicals. *Topics in Stereochemistry*, 1969, **4,** 1.

S. F. Nelsen. Nitrogen-Centred Radicals. In *Free Radicals* (ed. J. K. Kochi), vol. II, chapter 21. Wiley, New York, 1973.

E. G. Janzen, Stereochemistry of Nitroxides. *Topics in Stereochemistry*, 1971, **6,** 177.

W. G. Bentrude. Phosphorus Radicals. In *Free Radicals* (ed. J. K. Kochi), vol. II, chapter 22. Wiley, New York, 1973.

Chapter 6

A. R. Forrester, J. M. Hay and R. H. Thompson. *Organic Chemistry of Stable Free Radicals.* Academic Press, London, 1968.

C. Rüchardt. Relations Between Structure and Reactivity. In *Free-Radical Chemistry. Angewandte Chemie (International Edition)*, 1970, **9,** 830.

G. D. Mendenhall, D. Griller and K. U. Ingold. Prolonging the Life-Expectancy of Unconjugated Organic Free Radicals. *Chemistry in Britain*, 1975, **10,** 284.

M. Ballester. Inert Carbon Free Radicals. *Pure and Applied Chemistry*, 1967, **15,** 123.

D. Griller and K. U. Ingold. Persistent Carbon-Centred Radicals. *Accounts of Chemical Research*, 1976, **9,** 13.

Chapter 7

M. J. Gibian and R. C. Corkey. Organic Radical–Radical Reactions. Disproportionation *vs.* Combination. *Chemical Reviews*, 1973, **73**, 441.

T. Koenig and H. Fischer. Cage Effects. In *Free Radicals* (ed. J. K. Kochi),
vol. I, chapter 4. New York, 1973.

G. M. Burnett and H. W. Melville. Determination of Active Intermediates in
Reactions. In *Technique of Organic Chemistry* (eds. S. L. Friess, E. S.
Lewis and A. Weissberger), vol. VIII, part II, p. 1107. Wiley, New York,
1963.

Chapter 8

J. M. Tedder. The Interaction of Free Radicals with Saturated Aliphatic Compounds. *Quarterly Reviews*, 1960, **14**, 336.

G. A. Russell. Reactivity, Selectivity, and Polar Effects in Hydrogen Atom
Transfer Reactions. In *Free Radicals* (ed. J. K. Kochi), vol. 1, chapter 7.
Wiley, New York, 1973.

D. C. Nonhebel and J. C. Walton. *Free Radical Chemistry*, chapters 7–9.
Cambridge University Press, 1974.

Chapter 9

B. Lewis and G. von Elbe. *Combustion, Flames and Explosions of Gases.*
Academic Press, New York, 1961.

Chapter 10

C. Walling and E. S. Huyser. Free-Radical Additions to Olefins to Form
Carbon–Carbon Bonds. *Organic Reactions*, 1963, **13**, 91.

F. W. Stacey and J. F. Harris. Formation of Carbon–Hetero Atom Bonds by
Free-Radical Chain Additions to Carbon–Carbon Multiple Bonds. *Organic
Reactions*, 1963, **13**, 150.

J. M. Tedder and J. C. Walton. The Kinetics and Orientation of Free-Radical
Addition to Olefins, *Accounts of Chemical Research*, 1976, **9**, 183.

E. S. Huyser. *Free-Radical Chain Reactions*, chapter 8. Wiley, New York,
1969.

A. L. J. Beckwith. Some Aspects of Free-Radical Rearrangement Reactions.
In *Essays in Free-Radical Chemistry* (ed. R. O. C. Norman), p. 239.
Chemical Society Special Publication, No. 24, 1970.

M. Julia, Free-Radical Cyclizations, *Pure and Applied Chemistry*, 1967, **15**,
167.

Chapter 11

P. J. Flory. *Principles of Polymer Chemistry.* Cornell University Press, Ithaca, 1953.

G. E. Ham. *Vinyl Polymerization.* Part I of *Kinetics and Mechanisms of Polymerization*, Edward Arnold, London, 1967.

S. R. Sandler and W. Karo. *Polymer Syntheses*, vol. I. Academic Press, New York and London, 1974.

J. M. Tedder, A. Nechvatal, and A. H. Jubb. *Basic Organic Chemistry Part 5*, chapters 8 and 9. Wiley, London, 1975.

Chapter 12

G. H. Williams. *Homolytic Aromatic Substitution.* Pergamon, Oxford, 1960.

G. H. Williams. Homolytic Aromatic Arylation. In *Essays in Free-Radical Chemistry* (ed. R. O. C. Norman), p. 35. Chemical Society Special Publication No. 24, 1970.

D. H. Hey, Arylation of Aromatic Compounds, *Advances in Free-Radical Chemistry*, 1967, 2, 47.

R. A. Abramovitch. Intramolecular Free-Radical Aromatic Substitution, *Advances in Free-Radical Chemistry*, 1967, 2, 87.

R. O. C. Norman and G. K. Radda. Free-Radical Substitutions of Heteroaromatic Compounds. *Advances in Heterocyclic Chemistry*, 1963, 2, 131.

F. Minisci and O. Porta. Advances in Homolytic Substitution of Heteroaromatic Compounds. *Advances in Heterocyclic Chemistry*, 1974, 16, 123.

R. O. C. Norman. Free-Radical Alkylation. *Chemistry and Industry (London)*, 1973, 874.

C. D. Johnson. *The Hammett Equation.* Cambridge University Press, 1973.

Chapter 13

J. S. Littler. The Mechanisms of Oxidation of Organic Compounds with One-Equivalent Metal Ion Oxidants. In *Essays in Free-Radical Chemistry* (ed. R. O. C. Norman), p. 383. Chemical Society Special Publication No. 24, 1970.

J. S. Littler and D. C. Nonhebel. Free-Radical Reactions Involving Metal-Containing Species, and Related Processes in Free-Radical Reactions (ed. W. A. Waters), MTP International Review of Science, vol. 10, Series Two, 1976, p. 211.

J. K. Kochi. Oxidation and Reduction Reactions of Free Radicals and Metal Complexes. In *Free Radicals* (ed. J. K. Kochi), vol. II, chapter 11. Wiley, New York, 1973.

N. L. Weinberg (ed.). *Techniques of Electro-Organic Synthesis*, vols. I and II (especially chapters 3–8). Wiley, New York, 1974.

P. D. McDonald and G. A. Hamilton. Mechanisms of Phenolic Oxidative Coupling Reactions. In *Oxidation in Organic Chemistry*, Part B (ed. W. S. Trahanovsky), chapter 2. Academic Press, New York, 1973.

C. Walling. Fenton's Reagent Revisited. *Accounts of Chemical Research*, 1975, **8**, 125.

J. F. Bunnett. The Base-Catalysed Halogen Dance and Other Reactions of Aryl Halides. *Accounts of Chemical Research*, 1972, **5**, 139.

N. Kornblum. Substitution Reactions which Proceed via Radical Anion Intermediates, *Angewandte Chemie (International Edition)*, 1975, **14**, 734.

Chapter 14

J. A. Howard. Homogeneous Liquid-Phase Autoxidations. In *Free Radicals* (ed. J. K. Kochi), vol. II, chapter 12. Wiley, New York, 1973.

W. G. Lloyd. Autoxidations. In *Methods in Free-Radical Chemistry*, 1973, **4**, 1.

Chapter 15

J. W. Wilt. Free-Radical Rearrangements. In *Free Radicals* (ed. J. K. Kochi), vol. I, chapter 8. Wiley, New York, 1973.

R. Kh. Friedlina. Rearrangement of Radicals in Solution. *Advances in Free-Radical Chemistry*, 1965, **1**, 211.

T. S. Stevens and W. E. Watts. *Selected Molecular Rearrangements*, chapter 5. Van Nostrand Reinhold, London, 1973.

T. L. Gilchrist and R. C. Storr. *Organic Reactions and Orbital Symmetry*, chapter 7. Cambridge University Press, 1972.

K. Heusler and J. Kalvoda. Intramolecular Free-Radical Reactions. *Angewandte Chemie (International Edition)*, 1964, **3**, 525.

R. H. Hesse. The Barton Reaction, *Advances in Free-Radical Chemistry*, 1969, **3**, 83.

Chapter 16

W. A. Pryor (ed.). *Free Radicals in Biology*, vols. I–IV. Academic Press, New York, 1976–1978. (Especially vol. I, chapters 1, 2, 4 and 6, and vol. II, chapter 6.)

P. F. Knowles, D. Marsh and H. W. E. Rattle. *Magnetic Resonance of Biomolecules*, Wiley, London, 1976. (Especially chapters 6 and 8.)

O. H. Griffith and A. S. Waggoner. Nitroxide Free Radicals: Spin Labels for Probing Biomolecular Structure. *Accounts of Chemical Research*, 1969, **2**, 17.

L. J. Berliner. Applications of Spin Labelling to Structure–Conformation Studies of Enzymes. *Progress in Bio-organic Chemistry*, 1974, **3**, 1.

H. Dugas. Spin Labelled Nucleic Acids. *Accounts of Chemical Research*, 1977, **10**, 47.

W. T. Roubal. Spin Labelling with Nitroxide Compounds. *Progress in the Chemistry of Fats and Lipids*, 1972, **13**, 63.

L. A. Witting. The Interrelationship of Polyunsaturated Fatty Acids and Antioxidants *in vivo*. *Progress in the Chemistry of Fats and Lipids*, 1970, **9**, 517.

D. A. Forss. Odour and Flavour Compounds from Lipids. *Progress in the Chemistry of Fats and Lipids*, 1972, **13**, 181.

Chapter 17

K. U. Ingold and B. P. Roberts. *Free-Radical Substitution Reactions.* Wiley, New York, 1971.

R. A. Jackson. Group IVB Radical Reactions, *Advances in Free-Radical Chemistry*, 1969, **3**, 231.

Index

abstraction of halogen 78–80
abstraction of hydrogen 15, 17, 62–77
acetoxy radicals 104
acyl radicals 13, 104, 157, 162
acyloxy radicals, decarboxylation of 104
β-acyloxyalkyl radicals, rearrangement of
    161
acyloxylation 130
addition reactions,
    of alcohols 96
    of aldehydes 13, 96
    of group IV radicals 99
    of haloalkanes 7, 89–93, 95, 96
    of halogens 98
    of heterocyclic compounds 96
    of hydrogen halides 98, 106
    of hydroxyl radicals 145
    of phosphorus compounds 100
    of thiols 101
    addition reactions,
    orientation of 90–4
    reversibility of 75, 101–3
    stereochemistry of 104–6
additions,
    to alkenes 7, 13, 88–106, 145
    to alkynes 97, 99, 107
    to dienes 108
aldehydes,
    additions to alkenes of 13, 96
    autoxidation of 157
    decarbonylation of 161, 163
alkanes, halogenation of 62–78
alkenyl radicals, cyclization of 106
alkoxy radicals 68, 69, 103, 104, 154,
    165, 179
alkyl copper intermediates 136, 146
alkyl halides, electron-transfer substitution
    reactions of 147
alkyl radicals,
    abstraction by 69, 78, 79
    additions of 92
    combination of 55, 57

disproportionation of 55, 57
    persistent 53
    reaction with oxygen of 82, 86, 151
    structure of 37–41
alkylation,
    of arenes 130
    of heteroaromatics 133
alkylbenzenes, hydrogen abstraction from
    72, 127
alkylperoxy radicals 82, 151–6, 158, 176,
    178
allylic halogenation 76, 80
allylic radicals 42, 54, 176
amines, as antioxidants 155
aminyl radicals 43, 61
anodic oxidations 18, 139, 140
antioxidants 155
arachadonic acid 178
aromatic compounds, oxidation of 139
aryl halides, electron-transfer substitution
    reactions of 148
aryl radicals 54, 123
autoxidation,
    of aldehydes 157
    of alkanes 152
    of alkenes 158
    of lipids 153, 175, 178
    of tetrahydrofuran 152
azobisisobutyronitrile (AIBN) 12, 113
azo-compounds, decomposition of 8, 12,
    60

Barton reaction 166
benzoyloxy radicals 104, 130
benzyl radicals 46, 127, 132
Birch reduction 144
bond dissociation energies 1, 8, 10, 46, 65,
    151
brominations 75, 77, 92
N-bromosuccinimide 77
N-t-butyl α-phenylnitrone 28
t-butyl radicals 38

cage effects 12, 30, 59, 113, 168
carboxylic acids, oxidation of 18, 135, 137
chain transfer 62, 81, 116
chlorination, of alkanes 6, 62–4, 72–4, 80
    solvent effects in 77
CIDNP 30, 36, 168
coenzyme B_{12} 179–81
combination,
    of alkyl radicals 55, 57
    of alkylperoxy radicals 153
    of atoms 56
    of hetero radicals 60, 61
    of phenoxy radicals 141
    of polymer radicals 114, 115
conformations of radicals 25, 37–45
copolymerization 117–21
copper(II) salts, oxidation by 128, 136, 137
p-cresol, oxidation of 142
^{13}C splitting, in ESR spectroscopy 38
cumene hydroperoxide 150
cyclobutyl radicals 136, 137
cyclohexyl radicals 39, 130, 131, 190
cyclopropyl radicals 39, 164
cyclopropylmethyl radicals 106, 163
cytochrome a 174

diazonium salts 125, 129, 146, 147
dibenzoyl peroxide 13, 124, 188, 189
dimerization *see* combination
diphenylpicrylhydrazyl (DPPH) 52
disproportionation,
    of alkoxy radicals 154
    of alkyl radicals 55, 57
    of hetero radicals 60, 61
    of phenylcyclohexadienyl radicals 123, 124
    of polymer radicals 114, 115
di-t-butyl peroxide 12, 96, 189

electron-transfer reactions 10, 18, 130, 135, 136, 139, 140, 146, 147
electrophilic radicals 96, 127, 131
ESR spectra,
    of alkyl radicals 22–5, 37–41
    of allyl radicals 42
    of nitroxides 23, 31, 51, 182
    of phenoxy radicals 27
    of vinyl radicals 42
ethyl radical, ESR spectrum of 23
Evans–Polanyi equation 70, 71
explosions 84

Fenton's reaction 145
flavoproteins 172
fluoromethyl radicals 38

fragmentation of radicals 101
Fremy's salt 20, 23

galvinoxyl 7, 50
g-factors 21
glutathione 177
glycols, oxidation of 138
Gomberg reaction 124
group IV radicals 43, 66, 99

β-haloalkyl radicals 41, 75, 102, 161
halogen atoms,
    abstraction by 69
    abstraction of 78–80
halogenation of alkanes 13, 72
Hammett relationship,
    in homolytic aromatic substitution 131
    in radical halogenations 72
hex-5-en-1-yl radicals 107–9
homoalkyl radicals 106, 163
hydrazyl radicals 52
hydrogen, abstraction of 15, 17, 62–7
1,5-hydrogen migrations 164
hydroperoxides 104, 146, 150, 153, 154, 176–9
hydroperoxy radicals 85
hydroxyalkyl radicals 38
α-hydroxyethyl radicals 24
hydroxyl radicals 85, 131, 145
hydroxylation of aromatic compounds 131
hyperfine splitting 21–8, 38
hypophosphorous acid 147

iminoxy radicals 44
induced decomposition 13, 124, 188, 189
inhibition 147
initiators 13, 96, 110, 113

Jablonski diagram 15

Kaptein's rules 35
β-ketoalkyl radicals, rearrangement of 161
kinetics,
    of abstractions 66
    of additions 90
    of autoxidations 151
    of combinations 83
    of polymerization 110
Kolbe reaction 18, 135, 140

lead(IV), oxidation by 136–8
ligand-transfer oxidation 137, 146
lipids 176, 179, 184

McConnell equation 24
Meisenheimer rearrangement 169

mercury, as photosensitizer 16
methane, chlorination of 6, 64
methyl radicals 22, 38, 92, 130

NAD^+, oxidation with 171
neophyl radicals, rearrangement of 161
nitrosation of cyclohexane 80
N-nitrosoacetanilide 125
t-nitrosobutane 28
nitroxides 7, 23, 31, 44, 50, 169, 182
norbornenyl radicals 106, 163
nortricycyl radicals 163
nucleophilic radicals 131, 133

oxygen,
    m.o. structures of 5
    reactions with alkyl radicals 82, 86,
        151, 176

paramagnetism 4
partial rate factors in homolytic aromatic
        substitution 135
pent-4-en-1-yl radicals 107
peroxides, decomposition of 12
persistent radicals 6, 46, 53, 54
phenols,
    as antioxidants 155
    oxidation of 18, 135, 141
phenoxy radicals 18, 27, 49, 135, 141,
        175
phenylcyclohexadienyl radicals 123, 128
phenylethynyl radicals 130, 131
phosphinyl radicals 44, 61, 93, 100
phosphoranyl radicals 44, 187, 188
photolysis 13
polar effects in radical reactions 68, 70–2,
        79, 90, 93, 126, 127
polymer synthesis 121–3
polymerization, kinetics of 110–12
prostaglandins, biosynthesis of 178
Pschorr reaction 129, 146

radical anions 26, 145, 147–9
radical cations 139, 140
radical-pair theory 33–6
radiolysis 17
rates of chemical reactions,
    additions 90
    combinations 55, 57, 59
    disproportionation 58
    halogenation of alkanes 63
reductions 18, 135, 138
relative selectivities in abstractions 64, 69,
        70, 78
retention of configuration in radical reac-
        tions 37

Sandmeyer reaction 146
S_H2 reactions,
    at boron 185, 186
    at carbon 186, 191
    at group IV elements 187
    at magnesium 186
    at mercury 189, 190
    at oxygen 188, 189
    at phosphorus 185, 187, 188
solvent effects 2, 77
spin labels 182
spin polarization 24, 38
spin trapping 28
$S_{RN}1$ reactions 148
stable radicals *see* persistent radicals
stability of radicals 8, 46
stereochemistry,
    of radical additions 104–6
    of radical combinations 60
    of radical transfers 39, 43, 44
    of S_H2 reactions 190
steric effects,
    in abstractions 74
    in additions 93
Stevens rearrangement 168
styrene, autoxidation of 158
β-substituted alkyl radicals 39
superoxide 174

terminations 89, 114, 153
1,1,2,2-tetra-t-butylethyl radicals 53
thiyl radicals 61, 97, 101, 105
α-tocopherol 177
toluenes, halogenation of 92
trialkyltin radicals 80
triarylmethoxy radicals, rearrangement of
        162
triarylmethyl radicals 46
trihalomethyl radicals 17, 62, 69, 79,
        89–93, 95, 96, 102
triphenylmethyl radical 8, 18, 47
triplet state 4, 15
tri-t-butylmethyl radicals 38, 53
Trommsdorf–Norrish effect 116

ubiquinone 173
usnic acid 143

vinyl radicals 28, 42, 54
viscosity effects 59, 183
vitamin B_{12} *see* coenzyme B_{12}
vitamin E 177

Wittig reaction 168